AQA GCSE
(9-1) Combined Science: Trilogy

Required Practicals Lab Book

Emily Quinn

William Collins' dream of knowledge for all began with the publication of his first book in 1819. A self-educated mill worker, he not only enriched millions of lives, but also founded a flourishing publishing house. Today, staying true to this spirit, Collins books are packed with inspiration, innovation and practical expertise. They place you at the centre of a world of possibility and give you exactly what you need to explore it.

Collins. Freedom to teach

Published by Collins
An imprint of HarperCollins*Publishers*
The News Building
1 London Bridge Street
London SE1 9GF

Browse the complete Collins catalogue at
www.collins.co.uk

10 9 8 7 6 5 4 3

ISBN 978-0-00-829164-8

British Library Cataloguing in Publication Data
A catalogue record for this publication is available from the British Library.

Author: Emily Quinn
Commissioning editor: Rachael Harrison
In-house project editor: Isabelle Sinclair
Copyeditor and typesetter: Hugh Hillyard-Parker
Proofreader and answer checker: Stuart Lloyd and Tim Jackson
Artwork: QBS Learning
Cover designer: Julie Martin
Cover photo: © Xandpic/Shutterstock.com
Production controller: Tina Paul
Printed and bound by: Martins the Printers, Berwick-upon-Tweed

MIX
Paper from
responsible sources
FSC™ C007454

This book is produced from independently certified FSC™ Paper to ensure responsible forest management.

For more information visit: www.harpercollins.co.uk/green

The publishers will gladly receive any information enabling them to rectify any error or omission at the first opportunity.

Contents

Practical skills are at the heart of any science qualification. Your AQA GCSE Science course requires you to develop these skills through completing a series of required practicals, which you will then be tested on in your exams. This lab book will help you record the results of your practical work, and provide you with some guidance so that you get the most out of your time completing each practical.

Ensure you write down everything you can about your practical work – remember you can refer back to this book when you're revising!

Learning outcomes

This is a summary of what you should have accomplished by the end of each required practical.

Apparatus list

Your teacher will ensure that all the apparatus you need for the practical can be found in the classroom. You can use this list to check that you have everything you need to start your work.

Maths skills required

This is a good reminder of the skills you will need to master and practise on your science course, which will be tested in your exams. There are also questions included throughout that let you practise your maths skills.

Formulae

Any formulae you need to know to complete your practical work are shown here. A full list of formulae can be found on pages 97 and 98.

Safety notes

You should always be aware of safety when completing any practical work. This list will help you be aware of any common safety issues. Your teacher will advise on safety information for each practical, so pay attention.

Common mistakes

We've included some of the common mistakes people make during their practical work so that you can look out for them and hopefully avoid making the same mistakes.

Method

Always make sure you read every step of the method before you begin work. This will help you avoid mistakes and will give you an idea of what outcomes to look for as you complete each step.

Record your results

At the end of every method is place to record the outcomes of your work. Make sure you keep your notes clear and neat.

Check your understanding and Exam-style questions

For each practical, there are questions designed to check your understanding of the work you've just completed. There are also exam-style questions, which are included to help you prepare for questions about practical work in the exams. Some of these questions are designed to test your maths skills and to check your understanding of the apparatus and techniques that you've been using, as you'll be tested on these aspects of practical work in your exams.

Higher Tier

HT If you see this symbol next to a question, then it is designed for Higher Tier content only.

Teachers should always ensure they consult the latest CLEAPSS safety guidance before undertaking any practical work.

4.1.1.2 Microscopy

The work of Antonie van Leeuwenhoek was pivotal in developments in understanding the microscopic world. The compound microcope he invented was used to observe detail in objects that were far too small for the human eye to see. Your task is to make slides of everyday objects and view them under a light microscope at low and high resolution. You must include a magnification scale. You will draw a scientific sketch of your object at two different magnifications, add labels (if possible) and a include magnification scale.

Learning outcomes	Maths skills required	Formulae
• Use a light microscope to observe a range of different objects. • Record observations from a light microscope using simple diagrams. • Use a scale bar to identify the size of images from microscope slides.	• Calculate the size of the real object, the size of the image or the magnification. • Express answers in standard form (1×10^x). • Use prefixes centi-, milli-, micro- and nano-.	• magnification = $\dfrac{\text{size of image}}{\text{size of real object}}$

Apparatus list

- light microscope
- slides
- cover slips
- everyday objects – string, newspaper, paper, hair, plastic, insect wings, selection of prepared plant and animal slides
- forceps
- adhesive tape

Safety notes

- If using glass slides and cover slips, be careful as these are delicate and can shatter.

Common mistakes

- You might struggle drawing a scientific sketch. Draw the outlines of the structures you see and always include a scale. Don't add any shading.

- You might forget your units.
  Remember: 1 m = 100 cm = 1000 mm = 1 000 000 µm = 1 000 000 000 nm

- If using water, saline or a dye for your sample, you might end up with bubbles under your slide. Try lowering your cover sheet at an angle to push out the air bubbles (see **Figure 1**).

Figure 1

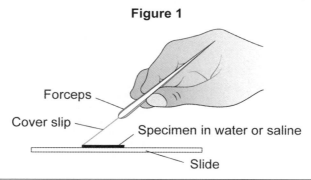

Forceps

Cover slip

Specimen in water or saline

Slide

Method

Read these instructions carefully before you start work.

1. Set up the microscope as you have been instructed by your teacher. Be careful with your microscope! It is one of the most expensive pieces of equipment in the laboratory.

2. Place the object on a microscope slide and place a cover slip or adhesive tape on top.

3. View the object under the microscope at low magnification (e.g. ×40).
   Draw your object as accurately as you can and label any structures you can see.

4. Now view your object at high magnification (e.g. ×400).
   Draw your object as accurately as you can and label any structures you can see.

5. Draw a suitable scale bar for your images.

6. Repeat steps **1**–**5** for as many objects as you have available.

Record your results

Draw one object at both low magnification and high magnification.

Draw another object at both low magnification and high magnification.

Check your understanding

1. Calculate the size of a real object under your microscope.

   Use the equation:

   $$\text{magnification} = \frac{\text{size of image}}{\text{size of real object}}$$
   [2 marks]

   ....

   ....

2. Explain why a scale bar is an important part of a microscopic drawing. [1 mark]

   ....

   ....

3. Explain why it is important to start at a lower magnification and then increase. [1 mark]

   ....

   ....

Exam-style questions

1. **a.** Humans have approximately 37 200 000 000 000 cells in the body.

    Express this number in standard form. [1 mark]

    ....

    b. The size of a human blood cell is 0.000 78 cm.

    Convert this from centimetres into micrometres. [1 mark]

    0.000 78 cm = ... μm

2. A student wants to observe and record the image of a plant root tip under a light microscope.

    a. Describe a method to observe and record the image of a plant root tip under a light microscope. [3 marks]

    ....

    ....

    ....

    ....

    ....

    ....

    b. Describe what the student should do to calculate the size of a cell in the plant root tip. [2 marks]

    ....

    ....

    ....

    ....

4.1.3.2 Osmosis

Osmosis is the diffusion of water through a partially permeable membrane. The water moves from a dilute solution to a concentrated solution. Plant tissues can be used to investigate osmosis. This experiment uses potato, but other tissues, such as sweet potato, carrot or beetroot, can be used. You are going to investigate the effect of a range of concentrations of salt solutions on the mass of plant tissue. You need use potato samples that are of equal size and place them in different salt solutions and distilled water. The changes in length and mass can then be accurately compared.

Learning outcomes	Maths skills required	Formula
• Measure changes in plant tissue due to osmosis. • Understand how different concentrations of solution affect plant tissue.	• Use ratios, fractions and percentages. • Make estimates of the results of simple calculations. • Plot two variables from experimental or other data. • Determine the slope and intercept of a linear graph.	• percentage mass change = $\dfrac{\text{change in mass of potato}}{\text{initial mass of potato}} \times 100$

Apparatus list

- potato chips
- sharp knife
- tile
- five Petri dishes with labels
- five different concentrations of salt solution
- electronic balance
- paper towels
- eye protection

Safety notes

- Be careful when using sharp knives!
- Be careful when using an electronic balance with water near by. Dry your potato samples before measuring their mass by patting them with a paper towel.

Common mistakes

- Do not squeeze the potato samples when you dry them. You are measuring the amount of water gained or lost. By squeezing the potato you will lose water and make your results invalid.

Method

Read these instructions carefully before you start work.

1. Collect five potato chips.
2. Place them on a tile and, using a sharp knife, trim them all to the same length.
3. Pour different concentrations of salt solution into five Petri dishes and label the dishes **1–5**, as shown in **Table 1**.
4. Measure the mass of each potato chip using the electronic balance and record it in **Starting mass in g** column of **Table 1**. Place each one in a separate Petri dish.
5. Repeat this with the other four dishes so that you have one chip in each dish.

6. Put covers over the dishes and leave for as long as possible but at least 20 minutes.

7. Remove the potato chips from the solution and place on a paper towel to remove the excess liquid.

8. Measure the mass of the chips again and record this in the **Final mass in g** column of **Table 1**. Calculate the change in mass by substracting the starting mass from the final mass.

9. Calculate the **Percentage (%) change in mass** by dividing the change in mass by the starting mass then multiplying by 100.

10. Plot the percentage change in mass against concentration of salt solution on **Graph 1**. Any results where the potato chip gained mass should be plotted above the 0% change in mass line, while any results where the potato chip lost mass should go below the line.

Record your results

Table 1 – Change in mass of potato samples

Petri dish	Concentration of salt solution in M	Starting mass in g	Final mass in g	Change in mass in g	Percentage (%) change in mass
1	0.0				
2	0.2				
3	0.4				
4	0.6				
5	0.8				

Graph 1

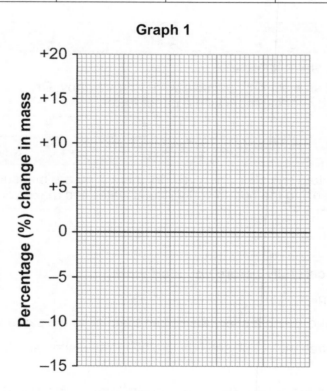

Check your understanding

1. Explain why is it important to put a lid on the Petri dish, especially if the sample is left for a long time. [2 marks]

   ....

   ....

   ....

2. Describe how this method could be changed to improve the accuracy of this experiment. [2 marks]

   ....

   ....

   ....

3. **HT** Osmosis occurs when there is a concentration gradient.

   Use your graph to estimate the concentration of the solution inside the potato cells. [1 mark]

   ....

Exam-style questions

1. A student investigates the effect of different salt solutions on potato tissue.

   The results are shown in **Table 2**.

   ### Table 2

Concentration of salt solution in M	Starting mass in g	Final mass in g	Change of mass in g	Percentage (%) change
0.0	1.40	1.62	0.22	15.7
0.2	1.45	1.61	0.16	11.0
0.4	1.40	1.46	0.06	
0.6	1.44	1.39	−0.05	−3.5
0.8	1.32	1.22	0.10	−7.6

   a. Calculate the percentage change in mass for 0.4 M salt solution. [2 marks]

   ....

   ....

   Percentage change in mass ....%

b. Explain why calculating percentage change of mass is a better method than only looking at change of mass. [2 marks]

....

....

....

c. The student suggests that the concentration of the solution inside the potato cell is between 0.4 M and 0.6 M.

Describe how the student can gain a more accurate estimate of the concentration of the solution inside the potato cell. [3 marks]

....

....

....

....

....

....

Foods can contain carbohydrates (starch and sugars), protein, lipids (fats) and small amounts of minerals and vitamins. Qualitative tests can be used to test for the presence of different food groups using ground-up food. The food is added to distilled water, stirred and filtered.

You are going to test some different foods for glucose (a simple sugar), protein, starch and lipids. This practical may take more than one lesson.

Learning outcomes	Maths skills required
• Use apparatus methodically when completing food tests. • Understand how to test for carbohydrates, lipids and proteins. • Interpret results to identify the types of substances present in food.	• Measure small volumes.

Apparatus list

• food: potato, carrot, crisps, biscuits, cheese • spotting tile • pipette • 10 cm^3 measuring cylinder or plastic syringe • beaker • boiling tubes • test tubes	• kettle • iodine solution • Benedict's reagent • biuret reagent • ethanol • eye protection

Safety notes

• Make sure you wear eye protection!
• Wash off spills on skin immediately.
• Do not eat or drink in the lab!
• Do not use ethanol around naked flames.

Common mistakes

• For the glucose test, make sure you heat the reagents for long enough.
• You only need a small amount of the reagents – do not use more than the volumes stated in the method.
• The colour change for the protein test is hard to determine – you are looking for blue changing to purple.

Method

Read these instructions carefully before you start work.

1. Choose a sample of food to test.

2. Carry out the four tests as described in **Table 1**.

Table 1 – Instructions for food tests

Food test	Steps
starch	1. Put a small piece of the food to test on a white tile. 2. Add two drops of iodine solution to the food. 3. If the iodine goes blue-black, the food contains starch.
glucose	1. Mix a small sample of the food with 3 cm^3 of Benedict's solution in a boiling tube. 2. Heat the mixture in a hot water bath for 3 minutes. 3. If the solution goes a tomato red colour, the food contains sugar.
protein	1. Mix a small sample of the food with 3 cm^3 of biuret solution. 2. Leave for 2 minutes. 3. If the mixture goes a pale purple colour, the food contains protein.
lipid	1. Mix a small sample of the food with about 1 cm^3 of ethanol in a dry test tube. 2. Pour the ethanol into a test tube full of cold water. 3. If the water goes milky white, the food contains lipid.

3. Record your observations in the row **1** of **Table 2**.

4. Repeat for four more food samples and record your observations in rows **2–5** of **Table 2**.

Record your results

Table 2 – Results of food tests

Food sample	Observations			
	Starch test	Glucose test	Protein test	Lipid test
1.				
2.				
3.				
4.				
5.				

Check your understanding

1. List two risks associated with this investigation. [2 marks]

   ....

   ....

2. Match up each reagent to the nutrient it tests for. [3 marks]

Benedict's		lipids
biuret		proteins
ethanol		glucose (a simple sugar)
iodine		starch (a complex carbohydrate)

3. This experiment required you to measure very small volumes of reagents.

   Suggest an appropriate piece of equipment you can use to measure very small volumes precisely. [1 mark]

   ....

Exam-style questions

1. A student tested three foods using qualitative food tests.

   The results are shown in **Table 3**.

   Table 3

Food sample	Observations			
	Starch test	**Glucose test**	**Protein test**	**Lipid test**
A	Brown	Blue	Purple	Milky white
B	Black	Tomato red	Blue	Colourless
C	Brown	Tomato red	Purple	Colourless

   For each of the food samples, list all the nutrients that are present.

   a. Food sample **A** .... [1 mark]

   b. Food sample **B** .... [1 mark]

   c. Food sample **C** .... [1 mark]

2. Sudan III is a reagent that can also be used to test for lipids.

   If lipids are present, a thin red layer is present floating on top of the sample.

   Suggest a food that would provide a positive test result with Sudan III. [1 mark]

   ....

4.2.2.1 Enzymes

The activity of enzymes can be affected by a number of factors such as temperature and pH. You are going to investigate the effect of pH on the rate of reaction of the enzyme amylase. This experiment involves a continuous sampling technique to determine the time taken to completely digest a starch solution at a range of pH values. Iodine reagent will be used to test for starch every 30 seconds. When starch is present, a blue-black colour will be seen when the iodine is added. If the starch has been broken down into sugar, then the iodine will remain orange-brown.

As pH is the independent variable being tested, temperature must be controlled. This can be done by using a water bath or immersible electric heater.

Learning outcomes	Maths skills required
• Use a water bath safely to control the temperature of an enzyme-controlled reaction. • Measure the rate of the reaction by the colour change of an iodine indicator. • Explain the effects of pH on the activity of amylase.	• Use ratios, fractions and percentages.

Apparatus list

• water bath (use a 250 cm^3 beaker containing 150 cm^3 of water kept at the temperature you require) • thermometer • 10 cm^3 plastic syringe or measuring cylinder • two 1 cm^3 plastic syringes • starch solution • amylase solution • iodine solution	• boiling tube • test tube • pipette • spotting tile • buffer solutions • stopwatch • eye protection

Safety notes

• Eye protection should be worn throughout.
• Take care with boiling water.

Common mistakes

• Always ensure you've set up correctly before you begin so that you can test the amylase starch mix every 30 seconds without delays.
• Avoid cross-contamination by making sure you use a different pipette for the different solutions.

Method

Read these instructions carefully before you start work.

1. Measure 10 cm³ of starch solution using the 10 cm³ plastic syringe and place this into the boiling tube.

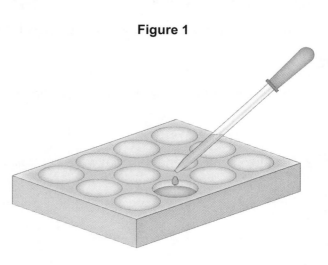

Figure 1

2. Measure 1 cm³ of a buffer solution using the 1 cm³ plastic syringe, then add this to the starch solution in the boiling tube.

3. Measure 1 cm³ of amylase solution using the other 1 cm³ plastic syringe, then add this to the test tube.

4. Place both the boiling tube and the test tube into the beaker of water to warm up.

5. Put one drop of iodine solution into each well of the spotting tile.

6. Add the amylase solution to the starch solution and mix. Start the stopwatch.

7. Immediately take out a drop of the starch/amylase mixture and add to a well in the spotting tile as shown in **Figure 1**. This is the sample for 0 seconds.

8. Repeat this every 30 seconds until there is no change in colour or 5 minutes have passed.

9. Repeat steps **1–8** for different pH values.

10. Record the results in **Table 1**.

Record your results

Table 1 – Effect of different pHs on the breakdown of starch by amylase

Time in seconds	Colour of iodine solution							
	pH.......	pH.......	pH.......	pH.......	pH.......	pH.......	pH.......	pH.......
0								
30								
60								
90								
120								
150								
180								
210								
240								
270								
300								

Check your understanding

1. Look at your results in **Table 1**.

   a. Which pH is the optimum value of pH for your amylase? [1 mark]

   ....

   b. At which pH values does your amylase not work? [1 mark]

   ....

2. List two control variables that were controlled during this experiment. [1 mark]

   ....

   ....

Exam-style questions

1. Lipase is a digestive enzyme that breaks lipids (fats) down into fatty acids and glycerol.
   Full fat milk has a pH of 6.5.

   a. Predict what will happen to the pH of full fat milk as lipids are broken down.

   Explain your prediction. [2 marks]

   ....

   ....

   b. A student wants to investigate the optimum pH of lipase.

   Describe a method the student could use to find the optimum pH of lipase. [6 marks]

   ....

   ....

   ....

   ....

   ....

   c. Lipase is an enzyme that works best in the alkaline conditions of the small intestine.

   Suggest a simple experiment to find the pH at which lipase works most effectively. [2 marks]

   ....

   ....

   ....

Photosynthesis is the process by which green plants produce food. When plants photosynthesise, they absorb light energy to power the reaction. They use carbon dioxide and water to produce glucose and oxygen.

The rate of photosynthesis is affected by many factors, such as:

- light intensity
- light wavelength.

You are going to investigate the effect of light intensity on the rate of photosynthesis using an aquatic organism such as pondweed. The effect of light intensity can be investigated by varying the distance between the pondweed and a light source. The closer the light source, the greater the light intensity.

Learning outcomes	Maths skills required
• Develop a hypothesis based on your understanding of photosynthesis. • Change one variable and observe the effect. • Control variables to ensure the data you collect is valid.	• Calculate the mean of a set of figures. • Understand inverse proportions **HT** • Present information graphically **HT**

Apparatus list

- test tube
- freshly cut 10 cm piece of pondweed (*Elodea* or *Cabomba* work well)
- light source
- ruler
- test tube rack
- stopwatch
- 0.2% solution of sodium hydrogencarbonate
- funnel
- eye protection

Safety notes

- Be careful with the lamps – they might get hot enough to burn your skin.
- Turn off any lamps when they are not in use.
- Clean up any spills immediately.
- Do not handle lamps, plugs, sockets or switches with wet hands and do not let water splash on the bulbs.

Common mistakes

- LEDs work better than lamps for this practical as they don't give out as much thermal energy. If you only have traditional lamps, put a large beaker of water in front of the hot light source; this will absorb the thermal energy but let the light pass through. If you do this, you may have difficulty getting the lamp close enough for your initial readings – see below!
- If there aren't any bubbles from the cut end of the pondweed, cut a few millimetres off the end and put it right next to the light source. It should then start bubbling and allow you to do your experiment.

Method

Read these instructions carefully before you start work.

1. Set up the equipment as shown in **Figure 1**.

2. Fill the beaker with water.

3. Place a fresh piece of pondweed into the bottom of the beaker and place a funnel over the top of the plant to hold it in place and help to collect the bubbles. Place your plant under high-intensity light to encourage photosynthesis.

4. Fill a test tube with sodium hydrogencarbonate solution and turn it upside down in the beaker. This is to make sure that there is sufficient carbon dioxide in the water so that carbon dioxide is not a limiting factor.

5. When the pondweed is producing a steady stream of bubbles, place it 10 cm from the light source and **start the stopwatch**. Place the test tube over the top of the funnel to make it easier to count the bubbles.

6. Count bubbles for 1 minute. **Stop the stopwatch** and record the results in the **Trial 1** column in **Table 1**.

7. Repeat twice more for 10 cm distance insert and record the results in the **Trial 2** and **Trial 3** columns in **Table 1**. Then calculate the mean bubbles per minute.

8. Repeat steps **5–7** with the pondweed at distances of 20 cm, 30 cm and 40 cm from the light source.

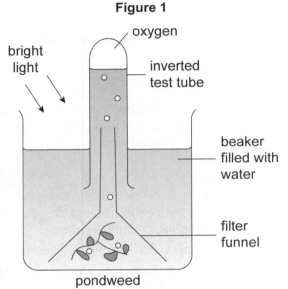

Figure 1

oxygen

bright light

inverted test tube

beaker filled with water

filter funnel

pondweed

Record your results

Table 1 – Effect of light intensity on rate of bubble production

Distance between pondweed and light source in cm	Number of bubbles per minute			
	Trial 1	Trial 2	Trial 3	Mean
10				
20				
30				
40				

Check your understanding

1. Consider the experiment.

   a. Write a hypothesis that predicts a link between light intensity and the rate of photosynthesis. [1 mark]

   ....

   b. Explain why you have made this hypothesis. [1 mark]

   ....

2. In the experiment, you controlled the amount of carbon dioxide available.

   State another variable that should be controlled and suggest **how** this variable could be controlled. [2 marks]

   ....

   ....

Exam-style questions

1. A student is investigating the effect of different wavelengths of light on photosynthesis. They use the following equipment:
   - a lamp
   - coloured light filters
   - a metre rule
   - a conical flask
   - 0.2% sodium hydrogencarbonate solution
   - a gas syringe
   - a sample of pondweed

   Plan an experiment to allow the student to collect valid data.

   You should identify any control variables. [6 marks]

   ....

   ....

   ....

   ....

   ....

   ....

   ....

   ....

   ....

2. **HT** In this experiment, sodium hydrogencarbonate solution is used to make sure carbon dioxide is not a limiting factor of photosynthesis.

State another factor that could be a limiting factor of photosynthesis in plants. [1 mark]

....

3. **HT** The relationship between the rate of photosynthesis and the distance of a plant from a light source is inversely proportional.

a. Explain what 'inversely proportional' means. [1 mark]

....

....

b. Draw a sketch graph to show the relationship between distance from a light source and rate of photosynthesis. [3 marks]

4.5.2.1 Reaction time

Your nervous system carries messages very quickly around your body to enable you to respond to changes in the environment. The amount of time it takes for you to respond to a change is called your reaction time.

There are different ways to improve your reaction time. For example, runners practise responding faster to the starting pistol. Also, some substances, such as caffeine, may improve reaction time.

You will conduct a simple, measurable experiment with a partner called the ruler drop test. From this you can determine whether your reaction time can be reduced.

Learning outcomes	Maths skills required
• Devise a hypothesis linking practice and reaction time. • Record and analyse data.	• Solve simple algebraic equations. • Translate information between numerical and graphical forms. • Plot two variables from experimental or other data.

Apparatus list	
• metre ruler • chair or stool	• eye protection

Safety notes
• Try not to hit yourself or your partner with the falling ruler.

Common mistakes
• Make sure you don't practise beforehand – this will make your results unreliable.

Methods

Read these instructions carefully before you start work.

1. Write a hypothesis in **Check your understanding**, **Question 1**, predicting how the amount of practice will affect the reaction time (for example, the more practice, the … the reaction time).
2. Use your non-dominant hand during this experiment. For example, if you use your right hand for writing, your non-dominant hand is your left hand and vice versa.
3. Sit on a chair or stool and place your non-dominant hand out in front of you. Your partner will stand and hold a ruler vertically with the bottom end (the end with the 0 cm) in between your thumb and first finger, as shown in **Figure 1**.

Figure 1

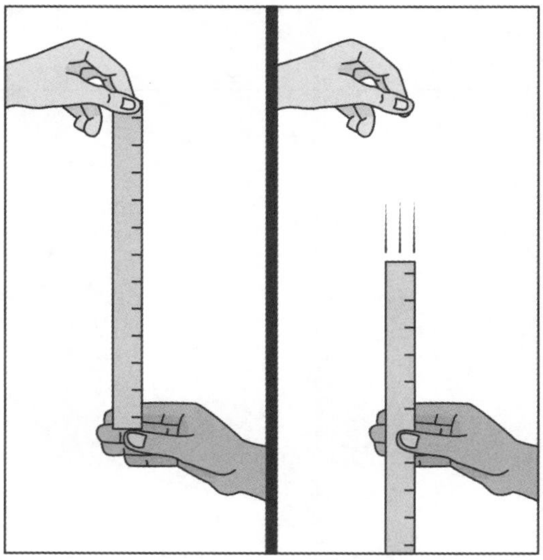

4. Your partner will drop the ruler **without telling you when**.

5. Catch the ruler as quickly as you can.

6. Read the number level with the top of your thumb. Record this in **Table 1**.

7. Rest for 30 seconds and then repeat steps **3–6** so you have 10 results.

8. Repeat the entire experiment with the roles reversed – that is, with you dropping the ruler and your partner catching it – in order to get your partner's reaction results.

9. Use **Table 2** to convert your ruler measurements into reaction times.

Record your results

Table 1 – Distance dropped and reaction time

Drop test attempts	Ruler measurements in cm		Reaction times in seconds	
	Person 1	Person 2	Person 1	Person 2
1				
2				
3				
4				
5				
6				
7				
8				
9				
10				

Reaction time in **Table 2** is calculated using the following formula:

$$t = \sqrt{(2d/g)}$$

where d = distance the ruler fell (m)

g = the acceleration of gravity (9.8 m/s²)

t = the time the ruler was falling (s)

Table 2 – Reaction time data

Distance (cm)	Reaction time (s)	Distance (cm)	Reaction time (s)	Distance (cm)	Reaction time (s)
1	0.05	34	0.26	67	0.37
2	0.06	35	0.27	68	0.37
3	0.08	36	0.27	69	0.38
4	0.09	37	0.27	70	0.38
5	0.10	38	0.28	71	0.38
6	0.11	39	0.28	72	0.38
7	0.12	40	0.29	73	0.39
8	0.13	41	0.29	74	0.39
9	0.14	42	0.29	75	0.39
10	0.14	43	0.30	76	0.39
11	0.15	44	0.30	77	0.40
12	0.16	45	0.30	78	0.40
13	0.16	46	0.31	79	0.40
14	0.17	47	0.31	80	0.40
15	0.17	48	0.31	81	0.41
16	0.18	49	0.32	82	0.41
17	0.19	50	0.32	83	0.41
18	0.19	51	0.32	84	0.41
19	0.20	52	0.33	85	0.42
20	0.20	53	0.33	86	0.42
21	0.21	54	0.33	87	0.42
22	0.21	55	0.34	88	0.42
23	0.22	56	0.34	89	0.43
24	0.22	57	0.34	90	0.43
25	0.23	58	0.34	91	0.43
26	0.23	59	0.35	92	0.43
27	0.23	60	0.35	93	0.44
28	0.24	61	0.35	94	0.44
29	0.24	62	0.36	95	0.44
30	0.25	63	0.36	96	0.44
31	0.25	64	0.36	97	0.44
32	0.26	65	0.36	98	0.45
33	0.26	66	0.37	99	0.45

Check your understanding

1. Write a hypothesis for this experiment, linking the amount of practice and the reaction time.

   [1 mark]

   ....

2. This is an experiment where it is difficult to take repeat readings using the same person.

   a. Explain why it is hard to take repeat readings using the same person. [1 mark]

   ....

   b. Suggest how you could improve the reliability of this experiment. [1 mark]

   ....

3. Caffeine is a stimulant that can be used to speed up reaction times.

   Plan a practical to find the effect of caffeine on reaction time. [2 marks]

   ....

   ....

Exam-style questions

1. A student planned an investigation into the effect of caffeine on reaction time.

   The student dropped a ruler for their partner to catch, first without caffeine, and then after having drunk a high-caffeine drink.

   The student measured how far the ruler dropped before their partner caught it.

   The results are in **Table 3**.

   Table 3

Test number	Experiment 1: student with no caffeine	Experiment 2: student after caffeine
	Distance the ruler dropped (mm)	
1	119	98
2	116	97
3	117	94
4	113	93
5	150	93
6	113	93
7	108	92
8	109	92
9	108	91
10	107	91

a. Describe any trends in the data.

Identify any anomalous results. [3 marks]

....

....

....

b. **Graph 1** shows a sketch graph of the data for **Experiment 2** given in **Table 3**.

Draw a sketch graph line on **Graph 1** for the **Experiment 1** data from **Table 3**.

Take into account any anomalous results. [1 mark]

Graph 1

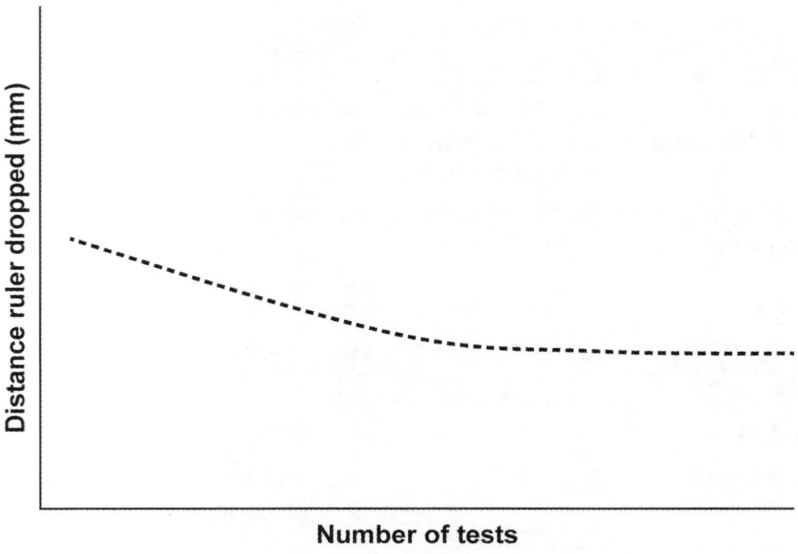

c. Suggest a reason for why the reaction time does not continue to decrease for **Experiment 2**. [1 mark]

....

It is impossible to count every plant or animal in a habitat, but total numbers can be estimated by taking samples of the organisms from the habitat. The larger the sample, the more accurate your estimate of the population size is likely to be. This allows population sizes to be compared between different areas.

Plants can be sampled more easily than animals because they cannot move around.

You are going to measure the population size of a common species in a habitat using random sampling. You will then use a different technique to investigate the effect of an abiotic factor on the distribution of this species.

There are two parts to this investigation:

1. measuring the population size of a plant species using random sampling

2. investigating the effect of a factor on that plant's distribution using a transect line.

Learning outcomes	Maths skills required	Formulae
<ul><li>Use results of an investigation to estimate the population.</li><li>Investigate the effects of an abiotic factor on population size.</li></ul>	<ul><li>Make estimates of the results of simple calculations.</li><li>Find arithmetic means.</li></ul>	<ul><li>estimated population size = mean population in 1 m^2 × total area (m^2)</li></ul>

Apparatus list

- 1 m by 1 m quadrat
- 2 × 30 m tape measures
- clipboard
- 10 pieces of paper, marked 1 to 10
- 10 pieces of paper, marked A to J
- two bags

Safety notes

- Be careful around plants: some can sting or cause allergic reactions. If you don't know what a particular plant is, don't touch it!
- Be careful when working in bushes or overgrown areas – there may be obstacles that you can't see.

Common mistakes

- Throwing a quadrat is not truly random as you are unlikely to throw it close to you and are limited by how far you can throw. It is also dangerous and you may break the quadrat.
- Don't move the quadrat to an area where there are more plants to count just to make things more interesting – your results will be invalid and you will have to repeat the experiment.
- Similarly, don't move the quadrat to where there are no plants so you have less to do – your results will be invalid and you will have to repeat the experiment.
- Depending on the time of year, there may be no flowers, only plant leaves. Ask your teacher for help with identifying plants by their leaves.

Methods

Read these instructions carefully before you start work.

There are two activities to complete.

Activity 1 – Estimating the population size of a plant species using random sampling

1. On the school field, look for a common plant that grows throughout the habitat such as dandelion.

2. Randomly place your quadrat by placing two tape measures alongside the '*x*-axis' and '*y*-axis' of the field.

    Divide the area into intervals the same size as your quadrat – for example, if you have a 1 m by 1 m quadrat, your intervals will be 1 m, 2 m, 3 m, etc.

Figure 1

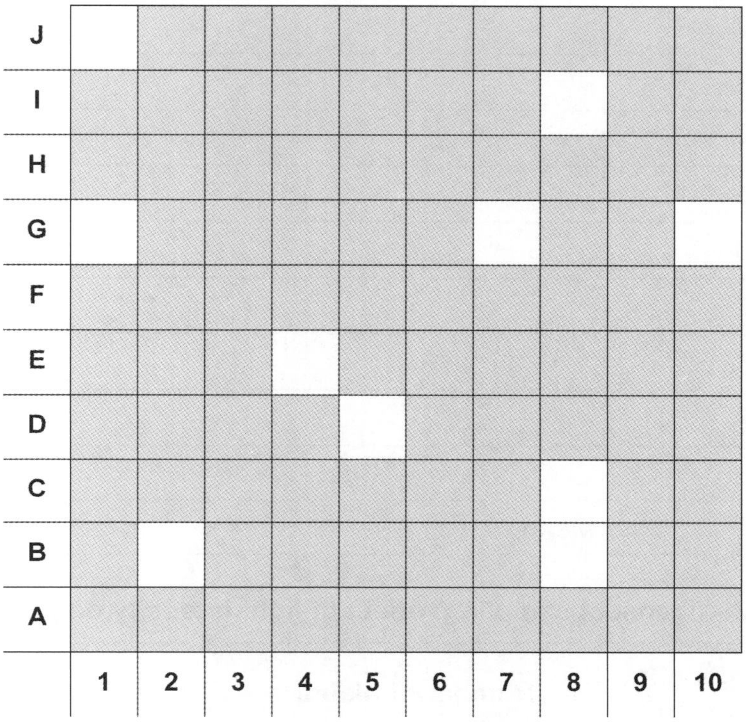

3. Place the numbers **1–10** in one bag and the letters **A–J** in another bag. Pull letters and numbers out of the bags to determine the ten areas you will sample. Replace the pieces of paper after each pair has been selected. Record your sample areas in **Table 1**.
    Figure 1 shows one result of such a random selection of ten quadrat positions.

4. Place your quadrat at each of the selected areas and record the number of the plants you are sampling in **Table 1**.

5. Calculate the mean number of plants per m² for your sample area.

6. Calculate the estimated population size for your sample area.

Activity 2 – Investigating the effect of an abiotic factor on that plant distribution using a transect line

1. On the school field, look for two areas where dandelions are growing, ideally under a tree starting in the shade and getting lighter as you move from under the tree.

2. Put down a transect line going from the shady area into the sunny area. Decide on the intervals at which you are going to place the quadrats. At least 10 samples should be taken – for example, for a 30 m transect, place a quadrat at 3 m intervals – 0 m, 3 m, 6 m, etc.

3. Place the quadrat down next to the line at the start. Use a light meter to measure the light intensity. Record the light intensity in **Table 2**.

4. Look inside the quadrat and count how many of the plants you are sampling there are. Record the number of plants in **Table 2**.

5. Repeat for each position along the transect line.

Record your results

Table 1 – Quadrat sampling: estimating population size

Quadrat number	Quadrat position	Number of plants
1		
2		
3		
4		
5		
6		
7		
8		
9		
10		
Mean		

Table 2 – Transect sampling: effect of light intensity on population

Distance along the transect line in m	Number of plants	Light intensity

Check your understanding

1. In **Activity 1**, you collected data for an area of 10 m^2 (10 × 1 m^2)
   (or a different area if you had a different-sized quadrat).

   Calculate the percentage of the total area that you sampled. [1 mark]

   ....

2. Describe what you could do to improve the accuracy of this experiment. [1 mark]

   ....

3. In **Activity 2**, your transect went from a shaded area to an area with higher light intensity.

   Describe any relationship between the number of plants and the light intensity. [1 mark]

   ....

4. Grass plants are too numerous to count different each individual plant.

   Explain how you could use a quadrat to gather data for a species like grass. [1 mark]

   ....

Exam-style questions

1. Some students want to investigate the relationship between which species are found on a tidal beach and the total time that the part of the beach is covered by sea water each day.

   They decide to use a transect.

   a. Suggest where the students should lay their transect. [1 mark]

   ....

   b. Describe a method that would allow the students to gather reliable data. [4 marks]

   ....

   ....

   ....

   ....

   ....

   ....

   ....

2. A student is surveying a population of daisies on the school field. The total area of the school field is 144 m². **Table 3** shows the results.

Table 3

Quadrat number	Number of plants
1	12
2	8
3	10
4	10
5	7
6	2
7	15
8	14
9	9
10	
Total	

Figure 2 shows the results for quadrat 10.

Figure 2

a. Fill in the result for Quadrat 10 in **Table 3**.   [1 mark]

b. Estimate the population of daisies in the field.   [2 marks]

   Use the equation:

   estimated population size = mean population in 1 m² × total area (m²)

   ....

   ....

Being able to choose the correct techniques and carry out a specified procedure to produce a pure product is an important skill for scientists. You will probably know how to complete each of the necessary techniques separately, but can you explain how to use them together to produce a product safely?

Your task is to prepare a pure, dry sample of a soluble salt from an insoluble oxide or carbonate, using a Bunsen burner to heat dilute acid and a water bath or electric heater to evaporate the solution.

You will react an acid and an insoluble oxide or carbonate to prepare an aqueous solution of a salt. The unreacted base from the reaction will need to be filtered. You will evaporate the filtrate to leave a concentrated solution of the salt, which will crystallise as it cools and evaporates further. When dry, the crystals will have a high purity.

Learning outcomes	Maths skills required	Formulae
• React a metal oxide with an acid to make a salt. • Work safely with hot chemicals. • Write word and chemical equations for this reaction.	• Measure precisely using a measuring cylinder • Discuss resolution of practical equipment	• M_r of a compound = A_r of all the elements present in the compound

Apparatus list

- 40 cm^3 1.0 M dilute sulfuric acid (M = mol/dm^3)
- 5 g copper (II) oxide powder
- spatula
- glass rod
- 100 cm^3 beaker
- Bunsen burner
- tripod
- gauze
- heatproof mat

- filter funnel
- filter paper
- 250 cm^3 beaker
- evaporating basin
- crystallising dish
- tongs
- eye protection

If crystallising dishes are not available, Petri dishes (without lids) make good substitutes.
If small conical flasks are not available, a second small beaker is an acceptable replacement.

Safety notes

- Wear eye protection at all times!
- **Do not** boil the acid! It can release harmful sulfur dioxide gas!
- Don't add large amounts of copper oxide in one go – the solution will boil and bubble over.
- If you allow the water bath to boil dry, it may cause the beaker to crack.
- You should not take the crystals home or pour copper sulfate down the sink!

Always be careful when handling chemicals and follow your teacher's safety advice about the below:

- 40 cm^3 1.0 M dilute sulfuric acid *(irritant)*
- copper (II) oxide powder *(harmful)*
- copper (II) sulfate *(harmful)*

- You might struggle getting the copper oxide to dissolve in the sulfuric acid. Add small amounts of copper oxide each time.
- You might struggle to heat the solution gently. Ask your teacher for a demonstration of how to heat using a gentle flame – with the air hole partly closed and the gas tap half open.
- If the solution starts to spit, move your Bunsen burner from underneath the solution.
- You might struggle to see where the crystallisation point is in step **6** below. You are looking for tiny crystals forming on the glass rod.

Method

Read these instructions carefully before you start work.

1. Pour 40 cm^3 of 1.0 M sulfuric acid into a 100 cm^3 beaker.

2. Set up your Bunsen burner on a heatproof mat with a tripod and gauze over it. Place the beaker with the acid on the gauze and heat gently over a gentle flame until almost boiling. **Turn the Bunsen burner off**.

3. Place the hot beaker on the heatproof mat using tongs. Add the copper oxide one spatula at a time, stirring using the glass rod as you add it. The mixture will turn clear and blue.

4. Fold and place a filter paper inside a funnel and filter the blue copper sulfate solution into a conical flask.

Figure 1

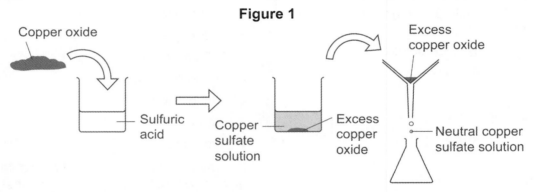

5. Pour the copper sulfate solution into the evaporating basin and heat over a 250 cm^3 beaker acting as a water bath until about half the solution has evaporated.

6. Test the solution by dipping a clean glass rod into it and then letting the rod cool. When small crystals form on the glass rod, stop heating the solution.

7. Use tongs to pour the copper sulfate solution into a crystallising dish and leave it in a warm place to finish crystallising.

Record your results

Write the word equation for this reaction.

....

Write the chemical equation for this reaction.

....

Describe the properties of the reactants and products, i.e. colour, state etc., in the **Tables 1** and **2** below.

Table 1 – Descriptions of reactants

Reactant	Chemical formula	Properties
Sulfuric acid		
Copper (II) oxide		

Table 2 – Descriptions of products

Product	Chemical formula	Properties
Copper sulfate crystals		
Water		

Check your understanding

1. Copper oxide is used in excess in the reaction with sulfuric acid.

   a. State what 'in excess' means. [1 mark]

   ....

   ....

   Copper oxide is used in excess to ensure this practical is safe.

   b. Describe what could happen if copper oxide was not used in excess. [1 mark]

   ....

2. Copper oxide (CuO) is reacted with sulfuric acid (H_2SO_4) to make copper sulfate ($CuSO_4$) and water (H_2O).

   a. Calculate the total relative formula mass (M_r) for the reactants of this reaction. [3 marks]

   ....

   ....

   ....

   b. State the law of conservation of mass. [1 mark]

   ....

c. Explain how the law of the conservation of mass allows you to predict the total M_r of the products.

Use this law to predict of the M_r of the products. [2 marks]

....

....

Exam-style questions

1. Copper oxide reacts with hydrochloric acid to form crystals of a new substance.

   a. State the name of the new substance formed. [1 mark]

   ....

   b. Suggest a method for obtaining a pure sample of crystals of this new substance. [6 marks]

   ....

   ....

   ....

   ....

   ....

   ....

2. A student reacts calcium carbonate with hydrochloric acid.

   a. Complete the symbol equation for this reaction: [1 mark]

   $CaCO_3 + HCl + \rightarrow CaCl_2 + H_2O +$ ....

   b. Name the salt made in this reaction.

   [1 mark]

   ....

3. Another way to make a salt is by adding a metal to an acid.

   a. Suggest the reactants needed to make the salt sodium nitrate. [1 mark]

   ....

   b. Suggest why this reaction would not be safe to do in the lab. [1 mark]

   ....

Electrolysis uses electricity ('electro-') to split ('-lysis') compounds in two. You are going to investigate what happens when aqueous solutions are electrolysed using inert electrodes and find out what products are made when these aqueous compounds are split using electricity. You will develop a hypothesis using your knowledge of reduction and oxidation and the reactivity series.

All the solutions contain H^+ ions and OH^- ions from partially ionised water molecules as well as the metal ions and non-metal ions from the soluble salt. For each electrolyte (the substance you perform electrolysis on) you will predict which element will be produced at the cathode and the anode and then see if your observations support your predictions.

Learning outcomes	Maths skills required
<ul><li>Safely carry out electrolysis.</li><li>Identify elements produced from your observation.</li><li>Apply your knowledge of oxidation and reduction.</li></ul>	<ul><li>Balance half equations.</li></ul>

Apparatus list	
<ul><li>copper (II) chloride solution</li><li>copper (II) sulfate solution</li><li>sodium chloride solution</li><li>sodium sulfate solution</li><li>100 cm³ beaker</li><li>Petri dish lid</li></ul>	<ul><li>two carbon rod electrodes</li><li>two crocodile clips</li><li>two leads</li><li>6 V power supply</li><li>blue litmus paper</li><li>tweezers</li></ul>

Safety notes
<ul><li>Wear eye protection at all times!</li><li>Chlorine is produced during two of the experiments. **Do not inhale it!** Chlorine gas (Cl_2) reacts with water (H_2O) on the mucous membranes in the lungs and produces hydrochloric acid (HCl), which then damages tissue in the lungs.</li><li>Make sure the laboratory is well ventilated (open the windows and put the extraction fan on).</li><li>**Do not** have a potential difference of higher than 4 V – too much chlorine will be made.</li><li>**Do not** run the electrolysis for more than 5 minutes – too much chlorine will be made.</li><li>If any of the equipment is damaged, **do not** use it.</li></ul>

Always be careful when handling chemicals and follow your teacher's safety advice about the below:

- copper (II) chloride solution *(harmful)*
- copper (II) sulfate solution *(harmful)*
- sodium chloride solution *(low hazard)*
- sodium sulfate solution *(low hazard)*

- Depending on the substance in the solution, other substances may be formed:
  - Gas produced at the positive electrode that does **not** bleach blue litmus paper is oxygen. If you collect the oxygen and add a glowing splint to it, it will relight it.
  - If a gas is produced at the negative electrode, it is hydrogen. If you collect the hydrogen and add a lit splint to it, it will make a pop.

Method

Read these instructions carefully before you start work.

1. Measure 50 cm^3 of copper (II) chloride solution and pour into the beaker.

2. Place the lid with carbon rods over the beaker OR attach the carbon rods to the edge of the beaker using crocodile clips. **The rods must not touch each other**.

3. Attach leads to the rods/crocodile clips and connect the rods to the **direct current (red and black) terminals** of a power pack (see **Figure 1**).

4. Set the potential difference at 4 V on the power pack and switch on for **5 minutes maximum**. You may see bubbles appearing very quickly.

Figure 1

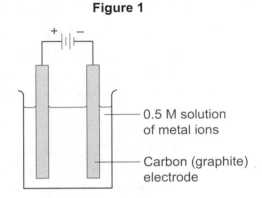

- 0.5 M solution of metal ions
- Carbon (graphite) electrode

5. After 30 seconds, observe the electrodes. You may see a gas being produced or a deposit being left on the electrode. Record your observations in **Table 1**.

   If a gas is produced, test it by holding a piece of damp litmus paper in it with tweezers.

   If the paper is bleached, you have a positive identification of an element. Record the result in **Table 1**.

6. After five minutes, **switch off the power supply.**

   Examine the surface of the electrodes for a coating. If there is a coppery-red deposit, you have a positive identification of an element. Record your results in the table.

7. Clean the equipment making sure there are no deposits left on the electrodes. Rinse with **distilled** water, not tap water.

8. Repeat steps **1–7** using:
   - copper (II) sulfate
   - sodium chloride
   - sodium sulfate.

Record your results

Table 1 – Products of electrolysis

Solution	Positive electrode (anode)		Negative electrode (cathode)	
	Observations	Element formed	Observations	Element formed
Copper (II) chloride				
Copper (II) sulfate				
Sodium chloride				
Sodium sulfate				

Check your understanding

1. Copper (II) chloride can be electrolysed to produce a copper deposit and chlorine gas.

   At which electrode will copper collect? ....

   At which electrode will chlorine gas be produced? ....

   [1 mark]

2. During electrolysis, copper (II) chloride forms Cu^{2+} ions and Cl^- ions.

   a. What is the formula of copper chloride? [1 mark]

   ....

   b. How many electrons does Cu^{2+} gain at its electrode? [1 mark]

   ....

   c. Is Cu^{2+} reduced or oxidised? [1 mark]

   ....

3. During electrolysis of sodium chloride, two gases are collected.

   a. Describe how a student could identify both of these gases. [2 marks]

   ....

   ....

   b. Describe any precautions the student should take to minimise the risks during this experiment. [2 marks]

   ....

   ....

Exam-style questions

1. **Table 2** shows the results of electrolysis for a range of compounds.

Table 2

Solution	Positive electrode (anode)		Negative electrode (cathode)	
	Observations	Element formed	Observations	Element formed
Copper (II) chloride	gas produced	chlorine	red/brown metal deposit on electrode	copper
Lithium chloride	gas produced	chlorine	gas produced	hydrogen
Sodium chloride	gas produced		gas produced	

    a. Name the elements formed during the electrolysis of sodium chloride, indicating the electrode at which each of them forms. [2 marks]

    Write your answers in **Table 2**.

    b. **HT** Predict the products of the electrolysis of potassium iodide.

    Explain why you have made this prediction. [4 marks]

    ....

    ....

    ....

    c. Predict the name of the solution left in the water. [1 mark]

    ....

2. During the electrolysis of aqueous copper (II) chloride, copper metal and chlorine gas are formed.

    a. **HT** Describe the reaction that occurs at the negative electrode that allows copper metal to be formed.

    Include a half equation for the reaction. [2 marks]

    ....

    ....

The electrolysis of molten lead bromide can be carried out using the equipment in **Figure 2**.

Figure 2

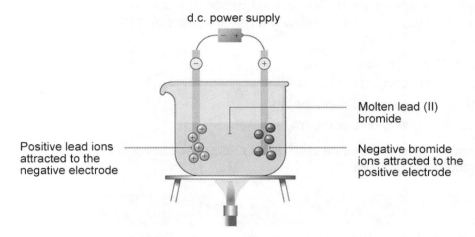

d.c. power supply

Positive lead ions attracted to the negative electrode

Molten lead (II) bromide

Negative bromide ions attracted to the positive electrode

b. Explain why the lead bromide must be molten. [2 marks]

....

....

c. Lead is produced at the negative electrode when molten lead bromide is used, but hydrogen is produced when aqueous lead bromide is used.

Explain why. [3 marks]

....

....

....

5.5.1.1 Temperature changes

Reactions can be endothermic (take in energy from the surroundings) or exothermic (give out energy to the surroundings). You will investigate a range of variables that affect temperature changes in reacting solutions such as:

1. neutralisation reactions
2. reactions between acids and carbonates
3. reactions between acids and metals
4. reactions involving displacement of metals.

You will monitor the temperature change as these reactions occur. The reaction will be contained in an insulated cup to reduce heat loss.

Learning outcomes	Maths skills required
• Carry out a range of temperature change reactions. • Be able to identify which changes are exothermic and which are endothermic.	• Translate information between graphical and numerical form.

Apparatus list

- expanded polystyrene cup and lid
- 250 cm³ beaker
- 10 cm³ measuring cylinder
- 50 cm³ measuring cylinder
- thermometer or temperature sensor
- eye protection
- spatula
- graph paper
- glass rod

- 1 M hydrochloric acid *(low hazard)* and 1 M sodium hydroxide solution (corrosive)
- 1 M ethanoic acid (low hazard) and a range of powdered metal carbonates (see CLEAPSS guidance)
- 0.5 M hydrochloric acid, strips of magnesium and pieces of zinc
- 1 M copper (II) sulfate solution and a range of metals

Safety notes

- Wear eye protection when doing all the practical activities and clearing up.
- Avoid skin contact with all the reactants.
- Some reactions can be **very** exothermic! **Do not touch** hot reactions and take care not to get burnt.

Always be careful when handling chemicals and follow your teacher's safety advice about the below:
- 1 M hydrochloric acid *(low hazard)*
- 1 M sodium hydroxide solution *(corrosive)*

Common mistakes

- 30 cm thermometers are easier to read than 15 cm. In this experiment, shorter thermometers would have most of the scale below the lid of the polystyrene cup so would be harder to read.
- You can use other material such as wood or cardboard if you don't have polystyrene cup lids, as long as the cup is kept covered to reduce heat loss by convection.
- Be sure to keep the bulb end of the thermometer in the water while taking measurements.

- When the reaction takes place, stir the mixture to distribute the heat evenly.
- Ask your teacher if you need help drawing two lines of best fit so that they intersect.

Methods

Read these instructions carefully before you start work.

There are four activities to complete.

Start by setting up the equipment as shown in **Figure 1** or in **Figure 2**.

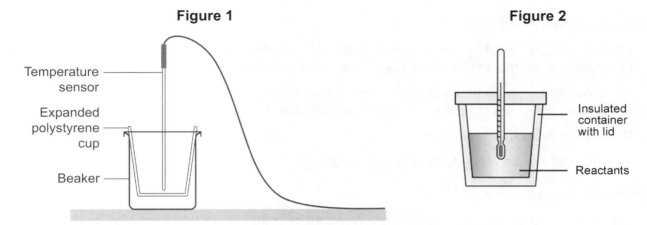

Figure 1

Figure 2

Activity 1 – Neutralisation

1. Measure 50 cm³ of hydrochloric acid using the larger measuring cylinder and pour into the polystyrene cup. Use a thermometer to measure the temperature of the acid. Record this in **Table 1**.

2. Measure out 5 cm³ of sodium hydroxide using the 10 cm³ measuring cylinder and carefully add the sodium hydroxide to the acid.

   Record the maximum temperature reached in **Table 1**.

3. Repeat step 3 adding another 5 cm³ until 40 cm³ of acid has been added.

   The last few additions should produce a temperature fall rather than a temperature rise.

4. Empty the cup, measure out 50 cm³ of fresh hydrochloric acid and repeat to gather a second set of data. Calculate the mean maximum temperature reached for each of the sodium hydroxide volumes.

5. On graph paper, plot a graph with:
   - **Mean maximum temperature in °C** on the *y*-axis
   - **Total volume of sodium hydroxide added in cm³** on the *x*-axis.

   Draw two straight lines of best fit:
   - one through the points that are increasing
   - one through the points that are decreasing.

   Ensure the two lines are extended so they cross each other.

6. Use the graph to estimate how much sodium hydroxide solution was needed to neutralise 25 cm³ of dilute hydrochloric acid.

Activity 2 – Acids + carbonates

1. Measure out 20 cm^3 of ethanoic acid and pour into the polystyrene cup. Use the thermometer to measure the temperature of the acid. Record this in **Table 2**.

2. Add one spatula of a metal carbonate to the ethanoic acid.

    Record the maximum temperature reached in **Table 2**.

3. Repeat step **2** until five spatulas of metal carbonate have been added.

4. Repeat to gather a second sat of data for a different carbonate. Compare the results for the two carbonates.

Activity 3 – Acids + metals

1. Measure out 20 cm^3 of hydrochloric acid and pour in to the polystyrene cup. Use the thermometer to measure the temperature of the acid. Record this in **Table 3**.

2. Add three small pieces of magnesium ribbon to the hydrochloric acid.
    Record the maximum temperature reached in **Table 3**. Repeat twice more and calculate a mean.

3. Repeat the experiment with zinc.

4. Explain your results with reference to the reactivity of metals.

Activity 4 – Displacement

1. Measure out 20 cm^3 of copper (II) sulfate and pour in to the polystyrene cup. Use the thermometer to measure the temperature of the copper (II) sulfate. Record this in **Table 4**.

2. You have a range of metals. Add a small amount of the first metal to the copper (II) sulfate. Record the maximum temperature reached. Empty out the cup into a bowl, measure out 20 cm^3 of fresh copper (II) sulfate. Repeat twice more and calculate a mean.

3. Repeat step **2** with the same amounts of other metal samples.

4. Explain your results with reference to the reactivity of metals.

Record your results

Table 1 – Neutralisation

Volume of sodium hydroxide added (cm^3)	Maximum temperature reached (°C)	
	1	2
0		
5		
10		
15		
20		
25		
30		
35		
40		

Table 2 – Acids + carbonates

Number of spatulas of carbonate added	Maximum temperature reached (°C)	
	Carbonate 1	Carbonate 2
0		
1		
2		
3		
4		
5		

Table 3 – Acids + metals

Metals	Maximum temperature reached (°C)			
	1	2	3	mean
Magnesium				
Zinc				

Table 4 – Displacement

Metals	Maximum temperature reached (°C)			
	1	2	3	mean

Check your understanding

1. A polystyrene cup is used to minimise thermal energy loss.

   a. Explain why it is important to take steps to minimise thermal energy loss. [1 mark]

   ....

   b. How else could thermal energy loss be minimised in this experiment? [1 mark]

   ....

2. The student wants to measure 5.0 cm^3 of sodium hydroxide

   a. What is the resolution of the measuring cylinder that the student has to use to accurately measure this amount? [1 mark]

   ....

b. Explain why a repeat reading is made during this experiment. [1 mark]

....

Exam-style questions

1. **Table 5** shows the results of a reaction between magnesium and hydrochloric acid.

Table 5

Concentration of HCl (M)	Time taken for magnesium to disappear (s)			
	1	2	3	mean
0.50	140	140	141	141
0.75	101	111	102	
1.00	72	71	73	72

a. Calculate the mean for the reaction using 0.75 M.
Take in to account any anomalous results. [2 marks]

....

....

b. The results for this 0.50 M and 1.00 M in this experiment are **precise**.

Explain how the results for these two concentrations show that the experiment is precise. [1 mark]

....

c. Describe the relationship between concentration of acid and time taken for the magnesium to disappear. [1 mark]

....

d. Explain this relationship in terms of collision theory. [2 marks]

....

....

2. **HT** Reactions of metals with acids are exothermic. Explain why this reaction is exothermic, using the idea of chemical bonds being made and broken. [2 marks]

....

....

Collision theory helps us to understand and predict the rate of reaction when certain conditions are changed. To react, particles need to collide; they need to be in contact with each other, and they also need to collide with enough energy to react – they need to be going fast enough. We can increase the probability that they will collide by increasing the concentration.

You are going to investigate how changes in concentration affect the rates of reactions. You will measure this by measuring the volume of a gas produced in one experiment and by observing a change in turbidity (cloudiness) in another experiment.

Before you start these experiments, make a hypothesis: how will concentration affect rate of reaction? Be sure to include an explanation based on collision theory to back up your prediction. Write your hypothesis in **Question 1a**.

Learning outcomes	Maths skills required
• Devise a hypothesis. • Measure volumes precisely to ensure your results are valid.	• Plot two variables from experimental data.

Apparatus list

Activity 1 – Measuring a change in turbidity

- 40 g/dm^3 sodium thiosulfate solution
- 2.0 M hydrochloric acid
- 10 cm^3 measuring cylinder
- 100 cm^3 measuring cylinder
- 100 cm^3 conical flask
- eye protection
- black cross on a sheet of paper
- stopclock

Activity 2 – Measuring a gas produced

- eye protection
- side arm flask and gas syringe

or

- conical flask (100 cm^3), with a single-holed rubber bung and delivery tube to fit the conical flask, trough or plastic washing-up bowl and upside down 100 cm^3 measuring cylinder filled with water
- 50 cm^3 measuring cylinder
- clamp stand, boss and clamp
- stopclock
- magnesium ribbon cut into 3 cm lengths
- dilute hydrochloric acid (2.0 M and 1.0 M).

Safety notes

- Wear eye protection at all times.
- Sulfur dioxide is released during the reaction. Make sure the lab is well ventilated.

Always be careful when handling chemicals and follow your teacher's safety advice about the below:

- 2.0 M hydrochloric acid *(irritant)*
- 1.0 M hydrochloric acid *(low hazard)*
- 40 g/dm^3 sodium thiosulfate solution *(low hazard)*

Methods

Read these instructions carefully before you start work.

There are two activities to complete.

Activity 1 – Measuring turbidity

1. Measure 10 cm³ sodium thiosulfate solution with a 10 cm³ measuring cylinder and pour into a conical flask.

    Measure 40 cm³ water with a 50 cm³ measuring cylinder and add it to the 10 cm³ of sodium thiosulfate in the conical flask. The diluted sodium thiosulfate will be at a concentration of 8 g/dm³.

    Place the conical flask on the black cross and wash the 10 cm³ measuring cylinder.

2. Measure 10 cm³ of dilute hydrochloric acid with the clean 10 cm³ measuring cylinder.

3. Start the stop clock **as soon as you add the acid to the conical flask**. Keep swirling the flask gently.

4. Observe the reaction by looking down through the top of the flask as shown in **Figure 1**.

    Stop the clock when you can no longer see the cross.

    Take care not to breathe in any sulfur dioxide fumes.

Figure 1

5. Record the time it takes for the cross to disappear in the **First trial** column of **Table 1**.

    Rinse all the equipment and do two more repeats for this concentration.

    Calculate the mean time in seconds.

6. Repeat steps **1–5** four times for the following concentrations:

    • 16 g/dm³ (20 cm³ sodium thiosulfate + 30 cm³ water)

    • 24 g/dm³ (30 cm³ sodium thiosulfate + 20 cm³ water)

    • 32 g/dm³ (40 cm³ sodium thiosulfate + 10 cm³ water)

    • 40 g/dm³ (50 cm³ sodium thiosulfate + no water)

7. Plot your data on **Graph 1** and draw a smooth curved line of best fit.

Activity 2 – Measuring the volume of a gas produced

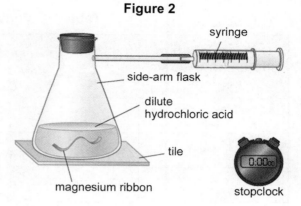

Figure 2

1. Set up the apparatus as shown in **Figure 2** <u>or</u> use a conical flask (100 cm³) and single-holed rubber bung and delivery tube to fit conical flask with an upside down 100 cm³ measuring cylinder filled with water in a troughor plastic washing up bowl.

2. Measure 50 cm³ of 2.0 M hydrochloric acid using a 50 cm³ measuring cylinder and pour the acid into the 100 cm³ conical flask.

3. Start the stop clock **as soon as you add the 3 cm strip of magnesium ribbon to the flask**, putting the bung back into the flask as quickly as you can.

4. Every 10 seconds, record the volume of hydrogen gas given off in **Table 2**. Continue timing until no more gas appears to be given off.

5. Repeat steps **1–4** using 1.0 M hydrochloric acid.

6. Plot your data for 2.0 M hydrochloric acid on **Graph 2** with:

    - **Volume of gas (cm³)** on the *y*-axis
    - **Time (s)** on the *x*-axis

    Draw a smooth curved line of best fit using a solid line.

7. Plot the points for 1.0 M hydrochloric acid on the same graph and draw a smooth curved line of best fit using a solid line using a dashed line or a different colour.

8. Use this graph to compare the rates of reaction of 1.0 M and 2.0 M hydrochloric acid with magnesium at 20 s. Do this by drawing a tangent to the 1.0 M curve at 20 s then calculate the gradient of the tangent. Repeat for the 2.0 M curve.

Record your results

Table 1 – Measuring a change in turbidity

Concentration of sodium thiosulfate in g/dm³	Time taken for cross to disappear (s)			
	First trial	Second trial	Third trial	Mean
8				
16				
24				
32				
40				

Graph 1

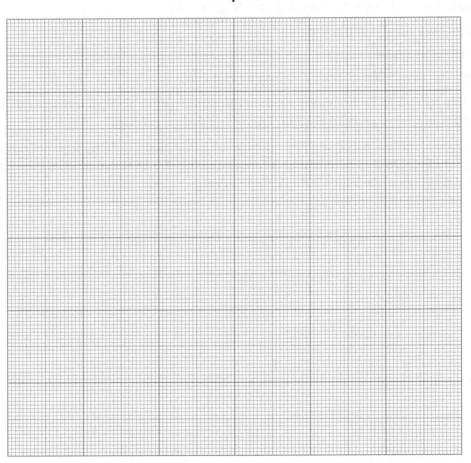

Table 2 – Measuring a gas produced

Time (s)	Volume of gas produced (cm^3)	
	2.0 M hydrochloric acid	1.0 M hydrochloric acid
10		
20		
30		
40		
50		
60		
70		
80		
90		
100		

Graph 2

Check your understanding

1. You are changing concentration in **Activity 1** and **Activity 2**.

   a. Write a **hypothesis** that predicts a link between concentration and time taken for a reaction to take place.

   Explain the link using collision theory. [2 marks]

   ....

   ....

   b. State another factor, other than concentration, that could be changed to affect rate of reaction. [1 mark]

   ....

2. In **Activity 2** you gathered one set of data for two different concentrations of hydrochloric acid.

   Describe how you could gather more data to support your hypothesis in **Question 1**. [1 mark]

   ....

Exam-style questions

1. Temperature affects the rate of chemical reactions.

   To see the effect of temperature on rate of reaction, sodium thiosulfate can be heated to different temperatures and then reacted with hydrochloric acid.

   a. Write a hypothesis that predicts a link between temperature and time taken for a reaction to take place.

   Explain the link using collision theory. [3 marks]

   ....

   ....

   b. Describe a method that could be used to find the rate of reaction between sodium thiosulfate and hydrochloric acid at different temperatures.

   List any equipment needed. [6 marks]

   ....

   ....

   ....

   ....

   ....

   ....

2. Hydrogen peroxide (H_2O_2) decomposes to form water and oxygen.

   a. Complete the equation for this reaction.

   Balance the equation. [2 marks]

   .... H_2O_2 → .... H_2O + ...

   b. The decomposition of hydrogen peroxide can be increased by using a catalyst.

   Explain why catalysts increase rates of reaction. [2 marks]

   ....

   ....

   c. Describe a method that could be used to measure how much a catalyst increases the rate of reaction of hydrogen peroxide decomposition. [2 marks]

   ....

   ....

Paper chromatography is a useful instrumental method to allow analysis of mixtures. You will investigate how paper chromatography can be used to separate and tell the difference between coloured substances.

You will calculate R_f (retention factor) values for the substances separated. The R_f value can never be higher than 1. R_f values that are close to 1 (e.g. 0.9) show that the food colouring pigments have spent more time in the mobile phase than in the stationary phase. Lower R_f values show the food colouring pigments have spent more time in the stationary phase than in the mobile phase. (If you are able to record this experiment using time lapse, it's very satisfying to watch when sped up).

Learning outcomes	Maths skills required	Formula
• Make and record measurements used in paper chromatography. • Calculate R_f values.	• Calculate R_f value. • Use ratios, fractions and percentages. • Substitute numerical values into algebraic equations. • Use appropriate units for physical quantities.	• $R_f = \dfrac{\text{distance moved by substance}}{\text{distance moved by solvent}}$

Apparatus list

- 250 cm^3 beaker with lid
- glass rod
- a rectangle of chromatography paper
- deionised water (solvent)
- glass capillary tubes
- four known food colourings labelled **A** to **D**
- an **unknown mixture** of food colourings labelled **U**
- paper clips
- hairdryer

Safety notes

- Try not to snap the capillary tube. They are made of thin glass.

Common mistakes

- Make sure you draw the line at the bottom of the chromatography paper in pencil. If it is drawn with ink, it will run and ruin the chromatogram.
- Do not let the food colouring spots touch the water; otherwise the food colouring will run into the water rather than up the chromatography paper.
- Don't let the chromatography paper touch the side of the beaker. If you do, it will distort the solvent front and make it harder to calculate an accurate R_f value.
- Don't forget about it! If the food colourings run off the edge of the chromatography paper, then you cannot calculate the R_f value, as you cannot measure the solvent front.
- Also, don't take the paper out too early. If the food colouring is moving with the solvent front (in the mobile phase), let it run a little further until it stops and is left behind on the paper (the stationary phase). If you take the paper out too early, all your food colourings will travel the same distance as the solvent front, which will make all your R_f values 1.

Methods

Read these instructions carefully before you start work.

1. Draw a pencil line 1.5 cm from the bottom edge of the chromatography paper.

2. Place a spot of the unknown substance on the pencil line using the capillary tube. Place spots of other food colourings alongside using different capillary tubes. The spots need to be small and concentrated. You can add more food colouring to each spot when it is dry. Record the order of the colours on the paper.

3. Add deionised water to the beaker so that it is 1 cm deep. Place the chromatography paper in the beaker so that the water can rise up the chromatogram (see **Figure 1**).

   The spots of food colouring **must** be above the water level. You can use the paper clips to hold the paper in place.

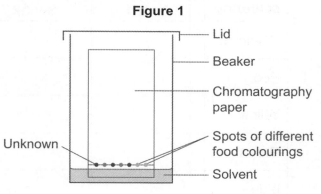

Figure 1

4. Allow the solvent (water) to rise up the chromatography paper until it is almost at the top. Remove the paper from the solvent and use a hair dryer to dry the chromatogram. Alternatively, you can leave them on a radiator.

5. Measure:
   - the distance from the pencil line to the solvent front
   - the distance from the pencil line to the middle of each spot.

   Your food colouring may separate out into a number of spots – you have space to record five spots. There may only be one or two spots. In this case, leave the other boxes blank.

   Record these values in **Table 1**.

Record your results

1. Distance/cm moved by the solvent = ...

Table 1 – Distance moved by food colourings

Food colouring	Distance moved by spot (cm)				
	Spot 1	Spot 2	Spot 3	Spot 4	Spot 5
Unknown					
Red					
Yellow					
Green					
Blue					

2. Calculate the R_f value for each spot on the chromatogram.

$$R_f \text{ value} = \frac{\text{distance moved by substance}}{\text{distance moved by solvent}}$$

(Each value should be between 0 and 1.)

Record these values in **Table 2**.

Table 2 – Calculating R_f values

Food colouring	R_f value of food colouring spots				
	Spot 1	Spot 2	Spot 3	Spot 4	Spot 5
Unknown					
Red					
Yellow					
Green					
Blue					

3. When dry, attach your chromatogram in the space below.

Check your understanding

1. A substance has an R_f value of 0.9.

   a. On the chromatogram, would you expect to see this substance the top near the solvent front **or** at the bottom near the pencil line? [1 mark]

   ....

b. Describe the solubility of this substance. [1 mark]

....

2. Another substance has an R_f value of 0.

  a. On the chromatogram, would you expect to see this substance the top near the solvent front **or** at the bottom near the pencil line? [1 mark]

....

  b. Describe the solubility of this substance. [1 mark]

....

3. Why is it important to add a lid to your chromatography beaker? [1 mark]

....

4. A student uses five different permanent markers for a chromatogram. She uses water as the solvent.

After running the chromatograph, the student calculates that **all** the permanent markers have an R_f value of 0.

Suggest two methods to improve her experiment. [2 marks]

....

....

Exam-style questions

1. Soy sauce is a mixture of different amino acids.

Thin layer chromatography is a technique that can be used to separate and identify amino acids.

The R_f values of different amino acids are listed in **Table 3**.

Table 3

Amino acid	R_f value
Alanine	0.30
Cystine	0.14
Phenylalanine	0.62
Serine	0.26

  a. Which amino acid spends the most time in the mobile phase? [1 mark]

....

  b. Which amino acid is most soluble in the solvent used in this experiment? [1 mark]

....

2. R_f values fall between 0 and 1.

a. Explain why an R_f value cannot be greater than 1. [1 mark]

....

Figure 2 shows a chromatogram of E numbers in food.

The R_f value of E131 is 0.9.

Figure 2

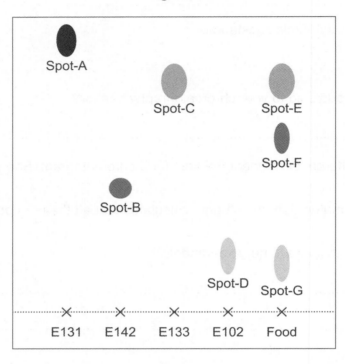

b. Calculate the R_f values for each of the spots in the food sample. [3 marks]

....

....

....

c. Which E numbers can you positively identify as being in the food sample? [2 marks]

....

....

d. Suggest how the experiment could be improved to identify the unknown sample of food. [3 marks]

....

....

....

Water that is safe to drink is called potable water. In a chemical sense, potable water is not pure water because it can contain dissolved substances that are needed by the body. Some countries do not have access to fresh water, only sea water. This has dissolved salts in it that need to be removed. This is done by distillation.

In this investigation you will test spring water, sea water and rain water for pH and the presence of dissolved salts. After distillation of the sea water, you will test the water again to check that dissolved salts have been removed. If they have, the water is fit for drinking! Don't actually drink it though! Nothing prepared in a school science lab is safe for human consumption. This practical may take more than one lesson.

Learning outcomes	Maths skills required
<ul><li>Safely purify water.</li><li>Compare pH and dissolved salts of a range of water samples.</li><li>Correctly use the apparatus, including the water bath.</li></ul>	<ul><li>Estimate of volumes</li><li>Calculate mass</li></ul>

Apparatus list

- 50 cm^3 sample of 'sea water'
- 10 cm^3 sample of 'spring water'
- 10 cm^3 sample of 'rain water'
- 10 cm^3 sample of 'sea water after distillation'
- universal indicator
- test tubes and rack
- Bunsen burner
- eye protection
- 10 cm^3 measuring cylinder
- 50 cm^3 measuring cylinder
- tripod
- gauze
- heatproof mat
- 250 cm^3 beaker
- watch glass
- tongs
- clamp stand
- 250 cm^3 conical flask
- delivery tube with bung
- anti-bumping granules
- ice

Safety notes

Activity 2

- Wear eye protection.
- **Do not** let the water bath boil dry!

Activity 3

- Wear eye protection.
- Make sure you have enough water in the conical flask to stop it boiling dry and cracking.
- Don't remove the conical flask from the heat during distillation! If the sea water starts to boil over, reduce the heat but do not remove it. This is to prevent suck back.
- Make sure you keep the delivery tube **at least 2 cm away** from the distilled water in the test tube. If you stop heating, the cold distilled water will be sucked back into the hot conical flask and can cause it to crack.

Methods

Read these instructions carefully before you start work.

There are three activities to complete.

(Your teacher might demonstrate the distillation of seawater to obtain water using a Liebig condenser rather than you doing **Activity 3**. You can use this distilled sea water as one of your samples rather than obtaining your own.)

Activity 1 – pH

1. Pour around 1 cm depth of

    a. sea water

    b. spring water

    c. rain water

    d. sea water after distillation

    into a test tube in the rack.

2. Add a one or two drops of universal indicator solution to each of the samples. Using a pH colour chart, match the colour and record the pH of the water in the **pH** column of **Table 1**.

Activity 2 – Collecting dissolved salts

1. Weigh a clean, dry watch glass. Record its mass in **Table 1**.

2. Set up your Bunsen burner on a heatproof mat with a tripod and gauze over it. Place the beaker with the water on the gauze (see **Figure 1**).

3. Measure 4 cm³ of one of your water samples using a 10 cm³ measuring cylinder and place it in a watch glass above a beaker with approximately 200 cm³ of tap water in it (see **Figure 1**). This beaker will act as a water bath to make sure your water sample will not boil too quickly.

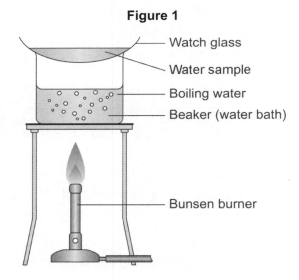

Figure 1

— Watch glass

— Water sample

— Boiling water

— Beaker (water bath)

— Bunsen burner

4. Heat the sample until all the water has evaporated from the watch glass and the dissolved salts are left behind. **Do not let the water bath boil dry**.

5. Remove the watch glass with tongs and allow to cool. Dry the underside of the watch glass with a paper towel to remove any water.

    Measure the mass of the watch glass with the salts and record the mass in **Table 1**. Calculate the mass of the dissolved solids. Wash the watch glass and dry it.

6. Repeat steps **3–5** for the other water samples. (If you use the same watch glass each time and you have cleaned it well, you do not need to measure the mass of the empty watch glass again.)

Activity 3 – Desalination of sea water

1. Pour the remaining salt solution into the conical flask. Add a few anti-bumping granules and set up the apparatus as in **Figure 2**.

2. Clamp the neck of the conical flask securely. Half-fill the beaker with ice and water.

3. Heat the salt solution with a blue Bunsen burner flame until it starts to boil. Adjust the flame so that the salt solution boils gently.

    Water will condense in the test tube. Collect about a 2 cm depth of distilled water.

4. Repeat **Activity 1** and **Activity 2** for your sample of distilled water

5. If you have been provided with different salt solution samples, you will need to repeat the procedures on these.

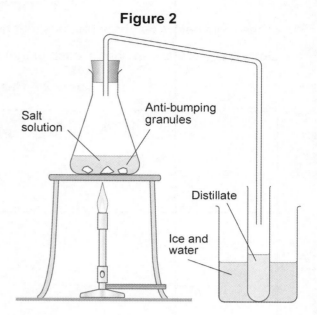

Figure 2

Salt solution

Anti-bumping granules

Distillate

Ice and water

Record your results

Table 1 – pH of sample and mass of dissolved solids

Water sample	pH	Mass (g)		
		Watch glass	Watch glass with dissolved solids	Dissolved solids
Sea				
Spring				
Rain				
Sea water after distillation				

Check your understanding

1. In the table below, list three hazards that need to be controlled during this experiment.

   Describe how you can control each of them. Write your answers in **Table 2**.   [6 marks]

Table 2 – Hazards and control methods

Hazard	Control method

2. The water provided for this experiment came from only one source.

   Explain how you could make your results a more accurate representation of the salts present in different sources of water from across the whole of the UK.   [3 marks]

   ....

   ....

   ....

3. Obtaining pure drinking water by the distillation of sea water is possible, but is rarely carried out.

   Suggest why distillation of sea water might be hard to achieve on a large enough scale to provide drinking water for a community.   [2 marks]

   ....

   ....

Exam-style questions

1. A sample of 10 cm³ of sea water is heated and evaporated in a watch glass to obtain the dissolved salts.

   The results are shown in **Table 3**.

   Table 3

Water sample	Mass (g)		
	Watch glass	**Watch glass with dissolved solids**	**Dissolved solids**
Sea	10.97	11.06	

   a. Calculate the mass of the dissolved solids.   [1 mark]

   ....

   b. Suggests how the accuracy of this experiment could be improved.   [2 marks]

   ....

   ....

2. Desalination is the process of removing excess salts from salt water or sea water to provide potable water.

   a. Describe a method for producing potable water from sea water.   [6 marks]

   ....

   ....

   ....

   ....

   ....

   ....

   b. Explain why potable water is **not** pure water.   [1 mark]

   ....

   ....

In this practical you will measure the temperature changes of different materials when they are heated so you can calculate their specific heat capacity. This investigation involves linking the decrease of one energy store (or work done) to the increase in thermal energy stored. The energy transferred (work done) will cause a temperature rise.

In this investigation, the temperature rise of a material depends on its **specific heat capacity**. Materials with a low specific heat capacity (a low capacity to store thermal energy) will have a greater temperature increase than those with a high specific heat capacity.

Learning outcomes	Maths skills required	Formulae
• Safely collect data to calculate the specific heat capacity of a metal block, or of water. • Understand how to identify any anomalous results.	• Substitute numerical values into algebraic equations using appropriate units for physical quantities. • Plot two variables from experimental data. • Determine the slope of a linear graph.	• change in thermal energy (J) = mass (kg) × specific heat capacity (J/kg °C) × change in temperature (°C) $\Delta E = mc\Delta\theta$ • energy transferred (J) = power (W) × time (s) • power (W) = potential difference (V) × current (A)

Apparatus list

- copper block with two holes, for a thermometer and a heater
- aluminium block with two holes, for a thermometer and a heater
- 250 cm³ beaker
- water
- thermometer
- petroleum jelly
- 50 W, 12 V heater
- 12 V power supply
- insulation to wrap around the block or beaker
- ammeter
- voltmeter
- five leads
- stopwatch
- balance (mass)
- eye protection

Safety notes

- Wear eye protection at all times.
- Be careful with water around electricity.
- The heating element will get very hot, especially if it is not inside a metal block. Take care not to burn yourself.
- If any of the equipment is damaged do not use it.
- If you scald yourself with hot water, cool the burn under cold running water immediately and ask your teacher for assistance.

Common mistakes

- The heating element should fit very snugly into the metal block, but there may be a small layer of air between the heating element and the metal block. Add a drop of water before you put the heating element in to improve transfer of energy between the heating element and the metal block.

- Remember to measure the mass of the metal block. These blocks are usually 1 kg, but to make sure your calculations are accurate, you should take an accurate mass measurement.

- Your teacher might tell you the power of your heater. You can trust your teacher, or you can attach an ammeter and voltmeter and calculate the power using:

  power (W) = potential difference (V) × current (A)
  $P = VI$

- Make sure you heat the metal block for at least 10 minutes; otherwise you will not be able to draw a graph with a good range of results.

- Don't forget to use your graph to find the gradient of the line. You will need this and the mass of the block to work out the specific heat capacity of your metals.

Method

Read these instructions carefully before you start work.

1. Choose your material, for example, a copper or aluminium block, and measure its mass, in kilograms. (Note: if choosing water, first find the mass of the beaker, then the mass of the beaker and the water, then subtract the mass of the beaker to determine the mass of the water.)

2. Wrap insulation around your block or beaker.

3. Smear petroleum jelly around the bulb end of the thermometer (not necessary if measuring temperature of water rather than of metal block) then put the thermometer in the small hole in the block (or into the water).

4. Measure the starting temperature of the block (or water).

Figure 1

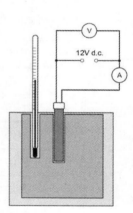

5. Put a heater in the larger hole in the block. Connect the ammeter, power pack and heater in series, as shown in **Figure 1**.

6. Connect the voltmeter across the power pack, as shown in **Figure 1**.

7. Turn the power pack to 12 V and switch it on. Start the stopwatch as you turn on the power pack.

8. Record the ammeter and voltmeter readings every 60 seconds in **Table 1**. These may vary slightly during the experiment, but not significantly. Record the temperature of the metal block (or water) every 60 seconds in **Table 2**. If you are using water, stir it at regular intervals.

9. After 10 minutes, turn off the power pack.

10. Keep the thermometer in the metal block (or water) for a while longer. Record the **maximum** temperature of the metal block (or water) in **Table 2** – this may be a little while after you have turned off the power pack.

11. Calculate the power of the heater and record the values in **Table 1**. Then calculate the energy transferred (work done) up to each time point. Use the following equations:
    power (W) = potential difference (V) × current (A)
    energy transferred (J) = power (W) × time (s)

12. Plot your results on **Graph 1**, with temperature change (°C) on the y-axis and energy transferred (J) on the x-axis.

Record your results

Table 1 – Calculating power

Time (s)	Potential difference (V)	Current (A)	Power (W)
0			
60			
120			
180			
240			
300			
360			
420			
480			
540			
600			

Table 2 – Calculating energy transferred

Time (s)	Energy transferred (J)	Temperature of metal block (°C)
0		
60		
120		
180		
240		
300		
360		
420		
480		
540		
600		
Maxiumum temperature of block		

Graph 1

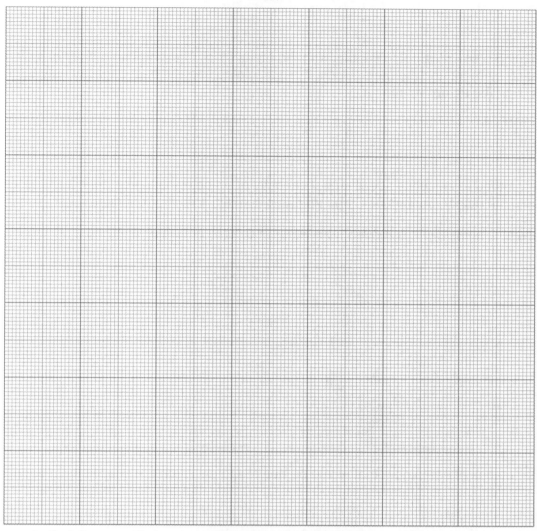

Theory suggests that the specific heat capacity can be found from:

$$\text{specific heat capacity} = \frac{1}{\text{mass} \times \text{gradient}}$$

Use your graph and this equation to determine the specific heat capacity of your metal.

....

Check your understanding

1. It is important to stir the water regularly when heating the water. Explain why.   [1 mark]

....

2. A student repeats the experiment for a block of copper four times.

   The results for the temperature of the block after 10 minutes are: 87 °C, 71 °C, 68 °C and 69 °C.

   a. Calculate the mean maximum temperature. Take into account any anomalous results.
   [2 marks]

....

....

b. Give one reason why the student may have anomalous results. [1 mark]

....

3. If the heating element is not in contact with the metal block, there will be a layer of air in between the heating element and the metal block. The calculated specific heat capacity may then be **greater** than the actual value.

Explain why. [1 mark]

....

4. Another method to calculate specific heat capacity of an object is to immerse the object in ice cold water until the object reaches 0 °C and then to place it in a beaker of hotter water.

In **Figure 2**, a brass block is moved from water at 0 °C to hot water at 80 °C.

Figure 2

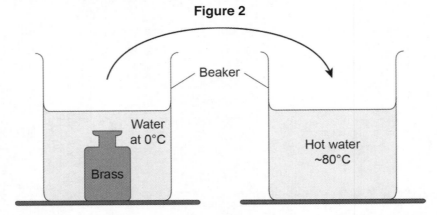

List three pieces of equipment that would be needed to calculate the specific heat capacity of the brass [3 marks]

....

....

Exam-style questions

1. 1 kg of aluminium was heated for 10 minutes.
   Table 3 shows the results.

Table 3

Energy transferred (J)	Temperature (°C)
0	23
4500	28
9000	33
13 500	38
18 000	43
22 500	48
27 000	53

a. Complete **Graph 2** using data from **Table 3**.

- Plot the additional data points.

- Draw a line of best fit. [3 marks]

Graph 2

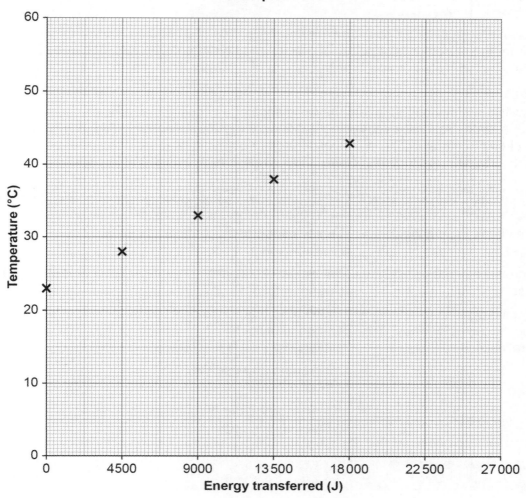

b. Determine the specific heat capacity of <u>this block</u> of aluminium. Use the gradient of the line and the formula below.

$$\text{specific heat capacity} = \frac{1}{\text{mass} \times \text{gradient}}$$
[2 marks]

....

....

2. The specific heat capacity for three different metals is listed in **Table 4**.

Table 4

Metal	Specific heat capacity (J/kg °C)
Aluminium	897
Copper	385
Iron	450

All three materials are commonly used as cooking pans. Which material would heat up the fastest? Give a reason for your answer. [2 marks]

....

....

We can use an electrical circuit to determine the resistance of a component, as resistance can be calculated from measurements of potential difference and current. We can then see how the resistance is affected by factors such as the length of a wire or when several components are combined.

A dimmer switch is a device that alters the resistance of a circuit and allows you to control the brightness of a lamp. You will investigate the principle of how a dimmer switch works. You will construct a circuit to measure the potential difference across a wire and the current through the wire. You will do this for different lengths of wire and for resistors in series and in parallel.

Learning outcomes	Maths skills required	Formulae
<ul><li>Set up a circuit to investigate resistance.</li><li>Make measurements using ammeters and voltmeters.</li><li>Use circuit diagrams to construct and check series and parallel circuits.</li></ul>	<ul><li>Display experimental data in a suitable table.</li><li>Plot two variables from experimental data.</li><li>Understand that $y = mx + c$ represents a linear relationship</li><li>Understand and use the symbol $\propto$</li><li>Use an appropriate number of significant figures.</li></ul>	<ul><li>$V = IR$</li><li>resistance = potential difference ÷ current</li></ul> $$R = \frac{V}{I}$$

Apparatus list

- d.c. power supply
- voltmeter
- ammeter
- length of resistance wire mounted on a metre rule

- connecting leads
- crocodile clips
- two 10 Ω resistors
- eye protection
- ohmmeter (if available)

Safety notes

- Short lengths of wire are likely to get hot. Use low values of potential difference. Switch off between readings.
- You can add a bulb to your circuit to stop the wire from getting too hot.

Common mistakes

- If your readings keep fluctuating, try to get an average value. Ammeters and voltmeters rarely stay at an exact value.

Methods

Read these instructions carefully before you start work.

There are two activities in this practical.

Activity 1 – Investigating how the length of a wire affects resistance

1. Set up the circuit as shown in **Figure 1**. The rectangle (symbol for a resistor) is where you should place the component that is being tested.

2. Attach your component, a metre rule with a wire, where the resistor symbol is. Use crocodile clips to add this component in to your circuit as shown in **Figure 2**.

Figure 1

Figure 2

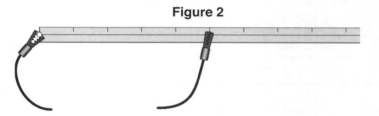

3. Place the crocodile clips so they are 100 cm apart.

4. Turn on the power pack at 4 V.

   Measure both the potential difference across the wire and the current through the wire. Record these results in **Table 1**.

5. **Turn off the power pack**.

   Reduce the length of the wire between the crocodile clips by 10 cm.

6. Repeat steps **4** and **5** for all wire lengths down to 10 cm.

   Be careful – the wire may get hot. Turn off the power pack if the wire starts glowing or smoking.

7. Calculate the resistance for each length of wire using the formula:

   resistance = potential difference ÷ current $R = \dfrac{V}{I}$

   Record the values of resistance in **Table 1**.

8. Plot a graph of resistance against length of wire in the space for **Graph 1**:

   • **Resistance in Ω** on the *y*-axis
   • **Length of wire in cm** on the *x*-axis.

Activity 2 – Investigating resistors in series and in parallel

1. Set up the circuit shown in **Figure 1** with two resistors in series where the rectangle is.

2. Turn on the power pack at 4 V.

3. Measure the potential difference across the power pack.

4. Measure the potential difference across each individual resistor and the current through them.

5. Calculate the total resistance of the circuit with resistors in series.

6. Set up the circuit for two resistors in parallel.

7. Repeat steps **2–4** for the resistors in parallel.

8. Calculate the total resistance of the circuit with resistors in parallel.

Record your results

Table 1 – Calculating resistance

Length of wire (cm)	Current (A)	Potential difference (V)	Resistance (Ω)
100			
90			
80			
70			
60			
50			
40			
30			
20			
10			

Graph 1

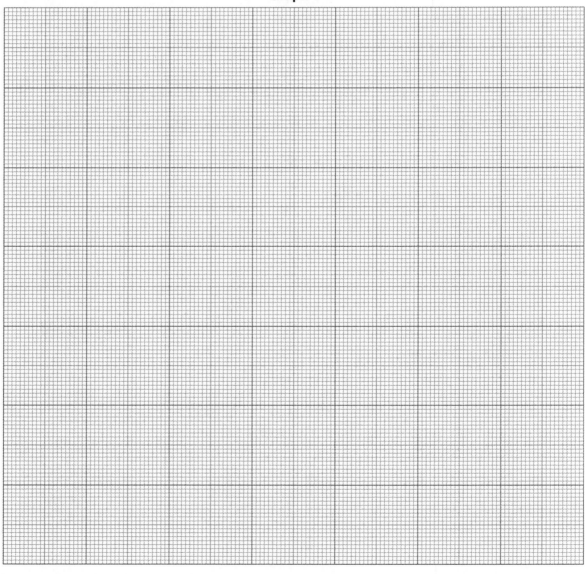

Activity 2 calculations

Check your understanding

1. Describe the relationship between the length of a wire and its resistance. [1 mark]

....

2. Use your graph to state whether this relationship is **proportional** or **directly proportional**.

   Justify your answer. [2 marks]

....

....

3. By adding a bulb to your circuit, you increase the resistance.

   Explain why this stops the wire from overheating. [2 marks]

....

....

Exam-style questions

1. State the units for potential difference, current and resistance. [2 marks]

   a. Potential difference (V) ....

   b. Current (I) ....

   c. Resistance (R) ....

2. State the equation that links potential difference, current and resistance. [1 mark]

....

3. Current is affected by the resistance of a wire.

   Describe what happens to the current as the resistance of a wire is increased. [1 mark]

....

Some components have a constant resistance. If we double the potential difference, the current doubles. We call these components **ohmic**. Other components do not have a constant resistance – increasing the potential difference might alter the current flow but it does not change proportionately. These are called **non-ohmic**. We can tell which are which by testing them, plotting graphs of the data and looking at the shape of the graph.

Learning outcomes	Maths skills required	Formulae
<ul><li>Use a range of circuits to safely gather data.</li><li>Make measurements using ammeters and voltmeters.</li><li>Use a graph to draw conclusions.</li><li>Identify variables that need to be kept constant in order to ensure validity of results</li></ul>	<ul><li>Plot two variables from experimental data.</li><li>Interpret graphs that represent a linear relationship.</li></ul>	<ul><li>potential difference (V) = current (A) × resistance (Ω)</li></ul>

Apparatus list

- 0–12 V variable power supply and connecting leads
- voltmeter or multimeter to measure V
- ammeter or multimeter to measure A
- milliammeter or multimeter to measure mA (you may need this if you have a diode only capable of small currents)
- variable resistor (for example, 10 Ω, 5 A)
- resistor (for example, 100 Ω, 1 W)

- filament lamp (for example, 12 V, 24 W)
- diode
- additional protective resistor (for example, 10 Ω) (optional) – you may need this to ensure your diode does not 'blow'; attach this in series with your diode if advised to by your teacher or science technician
- a switch to turn the circuit on and off (optional)

Safety notes

- Don't turn the power pack up too high! It will damage the components.
- The filament bulb and other components might get hot. Take care not to burn yourself on them.
- Check the equipment before use. If it appears damaged, don't use it.

Common mistakes

- Do not use a.c. in these experiments, use d.c. instead.
- It is **really** hard to set the potential difference to round numbers (e.g. 2 V, 3 V, 4 V) using the variable resistor. If you are struggling, you can take out the variable resistor and use the power pack values for potential difference – even though the potential difference may not be **exactly** what it says on the dial, the amount it increases by is the same each time you turn it up; the intervals are the same.

Diode problems – Activity 3

- If you are getting a current of 0 A or a very low current that doesn't increase at all, it might be that your leads are already switched for negative values. Try swapping them round.
- Don't forget the protective resistor if you have a low current diode! Without the protective resistor, you will blow the diode and then you can't get any results.

Methods

Read these instructions carefully before you start work.

There are three activities to complete.

Activity 1 – Resistor

1. Construct the circuit as shown in **Figure 1**.

2. For different settings of the variable resistor, record the values of current and potential difference. Take five pairs of readings. Turn off the powerpack in between readings to ensure the temperature of the resistor is kept constant.

3. Record the values in **Table 1**.

4. Swap the leads at the power supply (i.e. connect it so that the potential difference is negative).

5. Take readings as before and record them in **Table 1**.

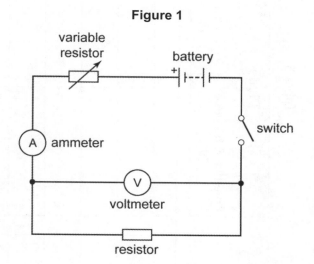

Figure 1

Activity 2 – Filament lamp

1. Construct the circuit as shown in **Figure 2**.

2. For different settings of the variable resistor, record the values of current and potential difference. Take at least five pairs of readings.

3. Record the values in **Table 2**.

4. Swap the leads at the power supply (i.e. connect it so that the potential difference is negative).

5. Take readings as before and record them in **Table 2**.

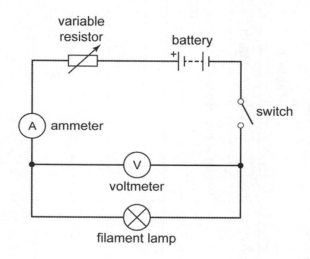

Figure 2

Activity 3 – Diodes

1. Construct the circuit as shown in **Figure 3** using an ammeter if you have a high-current diode, or a milliammeter if you have a low-current diode.

2. This activity may be tricky. Take any readings you can get, even if they are really close together. The range of possible results for this component can be tiny! After you have a few readings, try altering the setting on the variable resistor. Take four further pairs of readings, if possible.

3. Swap the leads at the power supply (i.e. connect it so that the potential difference is negative).

4. Take at least five pairs of readings as before.

5. Record all the values in **Table 3**.

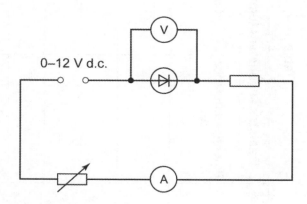

Figure 3

Plot graphs of current against potential difference for each component. Draw lines of best fit. Make sure you label your graphs clearly for each component.

Record your results

Table 1

Resistor at constant temperature

	Potential difference (V)	Current (A)
Positive values		
	0	0
Negative values		

Table 2

Filament bulb

Potential difference (V)	Current (A)
0	0

Table 3

Diode

Potential difference (V)	Current (A)
0	0

Check your understanding

1. Look at your graphs and describe the relationship between current and potential difference for a resistor at a constant temperature. [1 mark]

....

2. Explain why the power supply should be turned off between readings. [2 marks]

....

....

3. Calculate the resistance of your resistor at a constant temperature. [2 marks]

....

....

4. Describe how you can improve the accuracy of this calculated result. [2 marks]

....

....

Exam-style questions

1. When the potential difference across a filament lamp is 6.6 V, the current is 0.3 A.

   Calculate the resistance. Give the correct unit. [2 marks]

   ....

   ....

2. A student wants to investigate how the current through a diode affects its resistance.

   Describe how the student could use the circuit to investigate how current through a diode affects the resistance of the diode. [6 marks]

   ....

   ....

   ....

   ....

   ....

   ....

Scientists learnt how to measure density in an interesting way. Density is worked out from knowing the mass of an object and the volume it occupies. Measuring the volume is easy enough if the object is a regular shape such as a cube, but what if it was, say, a crown?

This problem was given to Archimedes, a clever scientist who lived thousands of years ago in Greece. The king had ordered a new crown to be made but suspected the craftsman had mixed a cheaper metal with the gold. Measuring the density would reveal whether the gold was pure or not, but how could the volume be found?

Archimedes realised that by carefully immersing the crown in a full can of water, the water that overflowed would have the same volume as the crown.

You are going to carry out an investigation into the density of different shaped objects – both regular and irregular – and of a liquid, gathering data for mass and volume.

Learning outcomes	Maths skills required	Formulae
• Calculate the density of regularly and irregularly shaped objects. • Take multiple readings to calculate a mean. • Measure length, volume and mass accurately using appropriate equipment.	• Calculate mean values from repeated measurements. • Calculate volumes of cuboids. • Record results to calculations using an appropriate number of significant figures.	• density = mass ÷ volume • volume of a cuboid = length × width × height

Apparatus list

- 50 cm^3 measuring cylinder
- 100 cm^3 measuring cylinder
- 250 cm^3 beaker
- water
- modelling clay

- balance (mass)
- a selection of regularly shaped rectangular objects
- ruler, micrometer or Vernier callipers

Safety notes

- Be careful when using water and the electronic balance.

Common mistakes

- In Activity 2, make sure your modelling clay object isn't too big, if the measuring cylinder overflows, you won't be able to measure the increased volume accurately.

- If your measuring cylinders are too small, your school might have Eureka cans that you can use. If you use Eureka cans, you only need to fill in column 1 – **Mass (g),** column 4 – **Volume of modelling clay (cm^3)** and column 5 – **Density (g/cm^3)** of **Table 2.**

Method

Read these instructions carefully before you start work.

There are three activities for this practical.

Activity 1 – Calculating the density of a regularly shaped object

1. Measure the length, width and height of a regularly shaped object using an appropriate piece of equipment for measuring. Record the dimensions in **Table 1**, using an appropriate number of significant figures.

2. Calculate the volume using the formula: volume = length × width × height

   Record the volume in **Table 1** using an appropriate number of significant figures.

3. Measure the mass of the object using the balance. Record the mass in **Table 1**.

4. Calculate the density using the formula: density = mass ÷ volume

   Record the density in **Table 1**.

5. Repeat for two more rectangular objects.

Activity 2 – Calculating the density of an irregularly shaped object

1. Fill the measuring cylinder to 50 cm^3 with water from the beaker. Record the exact volume in **Table 2**.

2. Take a blob of modelling clay and squeeze it into any shape you like, so long as it will fit into the measuring cylinder.

3. Measure and record the mass of your modelling clay shape in **Table 2**.

4. Carefully slide the modelling clay shape into the water by tilting the measuring cylinder at an angle. Make sure the modelling clay is totally covered.

5. Measure the new volume and record this in **Table 2**.

6. Calculate the volume change and record this in **Table 2**. This is the volume of the modelling clay shape.

7. Calculate the density of the modelling clay shape by using the formula: density = mass ÷ volume

   Record the density in **Table 2**, using an appropriate number of significant figures.

8. Repeat this twice more for different sizes of modelling clay shapes, also recording your results in **Table 2**.

9. Calculate a mean value for the density of the modelling clay.

Activity 3 – Calculating the density of a liquid

1. Measure the mass of an empty 50 cm^3 measuring cylinder. Record the mass in **Table 3**.

2. Fill the measuring cylinder to 50 cm^3 with water, or a different liquid. Record the exact volume in **Table 3**.

3. Measure the mass of the measuring cylinder and the liquid. Record the mass in **Table 3**.

4. Calculate the mass of the liquid by calculating the difference between your two measurments, and record the measurement in **Table 3**.

5. Calculate the density of the liquid by using the formula: density = mass ÷ volume.

Record your results

Table 1 – Density of regular objects

Object	Length (cm)	Width (cm)	Height (cm)	Volume (cm³)	Mass (g)	Density (g/cm³)

Table 2 – Density of irregular objects

Mass (g)	Volume without modelling clay (cm³)	Volume with modelling clay (cm³)	Volume of modelling clay (cm³)	Density (g/cm³)
Mean				

Table 3 – Density of a liquid

Mass of the empty cylinder (g)	Volume of liquid (cm³)	Masss of cylinder plus liquid (cm³)	Mass of liquid (g)	Density (g/cm³)

Check your understanding

1. In your investigation, you calculated the density of modelling clay.

    a. Explain why it was important to try more than one shape. [1 mark]

    ....

    b. Predict the effect of an air bubble trapped inside the modelling clay on the calculated density.

    Give a reason for your prediction. [2 marks]

    ....

    ....

2. A student has a regular cube that is too big to place inside a measuring cylinder.

   a. Suggest a piece of equipment that could be used to calculate the volume of the cube. [1 mark]

   ....

   b. Describe how this piece of equipment could be used to calculate the volume of the cube. [1 mark]

   ....

Exam-style questions

1. Petrol is a liquid with a density lower than that of water.

   Describe a method to calculate the density of petrol. [4 marks]

   ....

   ....

   ....

   ....

   ....

2. A 2.0 m³ block of aluminium has a mass of 5 400 000 g.

   Calculate the density of aluminium in kg/m³. [2 marks]

   ....

   ....

   Density of aluminium = ... kg/m³

3. A 2.0 m³ block of steel has a density of 7700 kg/m³.

   Calculate the mass of the steel block in kg. [2 marks]

   ....

   ....

   Mass of steel block = ... kg

4. Explain why aluminium is often used to build aeroplanes, but steel is not.

   Use your understanding of density to help you. [2 marks]

   ....

   ....

It is easy to make a spring by winding a length of wire around a cylindrical object. You can investigate how springs behave when loaded with a weight. You are going to investigate the relationship between the force applied to stretch a spring and its extension.

Learning outcomes	Maths skills required	Formulae
• Accurately measure the extension of a spring to calculate the spring constant. • Plot a graph of results and identify the limit of proportionality.	• Plot two variables from experimental data. • Interpret graphs that represent a linear relationship. • Calculate the gradient of a graph.	• force = spring constant × extension $F = ke$

Apparatus list

• spring	• ruler
• set of masses and mass holder appropriate for the spring you are testing	• 1 kg mass to put on base of clamp stand to stabilise it OR a table clamp
• clamp stand	• eye protection

Safety notes

• Be careful not to drop the masses on to your foot.

Common mistakes

• Don't confuse length with extension. The extension is the stretched length minus the original length of the spring.

• Try not to stretch the spring when you add and remove masses.

Method

Read these instructions carefully before you start work.

1. Set up the equipment as shown in **Figure 1**, but without any masses or the mass holder on the spring.

2. Measure the length of the spring with no masses attached. Record this length in **Table 1** in the **Length of spring (cm)** column for 0 N. The extension is filled in for you: no force means no extension.

3. Add a mass holder to the end of the spring and measure the length of the spring, using the pointer to improve accuracy.

4. Record the force on the spring and the new length of the spring in **Table 1**; a mass of 100g has a weight of 1N.

   Calculate how much the spring has extended:
   new length − original length for 0 N

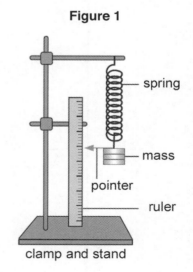

Figure 1

spring

mass

pointer

ruler

clamp and stand

5. Repeat steps **3** and **4**, adding the same amount of mass each time.

6. Stop when your spring shows signs of reaching the limit of proportionality. Make sure you do not overstretch your spring.

7. Plot a graph of your results for force against extension on **Graph 1**.

   Plot force on the *x*-axis and extension on the *y*-axis.

Record your results

Table 1 – Calculating extension

Force on spring (N)	Length of spring (cm)	Extension of spring (cm)
0		0

Graph 1

Check your understanding

1. Look at the graph and describe the relationship between force and extension. [2 marks]

....

....

2. Identify the limit of proportionality on your graph. [1 mark]

....

3. Suggest an improvement to this experiment that would improve the accuracy of your results. [1 mark]

....

....

4. Use $F = ke$ to calculate the spring constant of your spring. [1 mark]

....

....

Exam-style questions

1. **Graph 2** shows the results for an experiment investigating the extension of a spring.

Graph 2

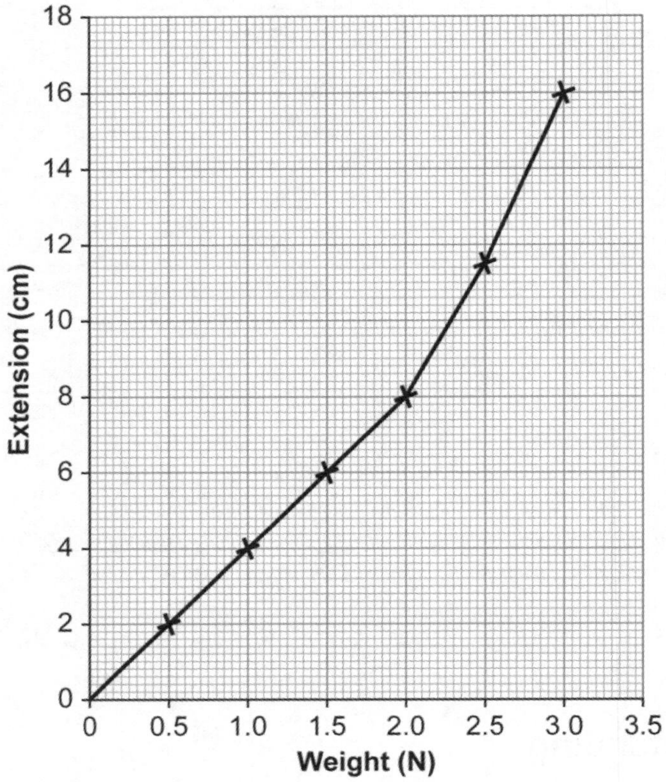

a. State the extension beyond which the limit of proportionality is exceeded. [1 mark]

....

b. State the formula linking force, extension and spring constant. [1 mark]

....

c. Use the graph to calculate the spring constant of the spring.

Give your answer in N/m. [3 marks]

....

....

....

....

....

d. The formula for calculating elastic potential energy is:

elastic potential energy = 0.5 × spring constant × extension^2

Calculate the amount of energy stored as elastic potential energy when the spring has 1 N of force applied to it.

Give the unit for your answer. [4 marks]

....

....

....

....

....

....

....

....

6.5.6.2.2 Acceleration

There is a very important relationship between force, mass and acceleration. This is a fundamental idea in Physics. It is possible to demonstrate the relationship using fairly straightforward apparatus and analysing the results with care.

You are going to carry out a practical investigation into Newton's second law.

Newton's second law of motion says that if a force accelerates an object, the acceleration is **directly proportional** to the force and **inversely proportional** to the mass of the object. You should be able to analyse your results to see whether they show this relationship.

Learning outcomes	Maths skills required	Formulae
• Make accurate measurements of length, mass and time. • Take multiple readings to calculate a mean. • Identify the independent and dependent variables, and the control variables.	• Find a mean. • Use an appropriate number of significant figures.	• $\text{speed} = \dfrac{\text{distance}}{\text{time}}$ • $\text{acceleration} = \dfrac{\text{change in velocity}}{\text{time taken}}$

Apparatus list

- light gate linked to a computer
- trolley with U-shaped interrupt card attached
- pulley on clamp
- string
- four 100 g masses and 100 g mass holder
- stopwatch
- box or tray to catch masses in
- pencil, chalk or masking tape to mark the intervals

Safety notes

- Be careful when masses are released from the bench as they might land on your feet.
- Prevent the trolley and interrupt card from falling off the bench.

Common mistakes

- Make sure you have your light gate and interrupt card at the right height. Ask your teacher for help if you don't know how to set them up.
- In Activity 1, make sure you don't change the mass of the system. Masses that aren't on the hook adding to the force accelerating the trolley need to be attached to/on the trolley. This makes sure that the total mass experiencing acceleration remains the same throughout your experiment.
- Don't forget that some mass hooks already have a mass of 100 g attached. Ask your teacher if you are unsure of the masses you have.
- For Method 2 in Activity 1, don't forget you will need to first calculate the average speed of the trolley in each segment. Then to find the acceleration of the trolley between segments find the change of speed from one segment to the next and the time the trolley took to speed up from one segment to the next.

Method

Read these instructions carefully before you start work.

There are two activities to complete.

Activity 1 – Measuring the effect of force on acceleration at constant mass

There are two different methods for this practical. Choose the method that best suits the apparatus available.

Method 1 – Using light gates

1. Set up the equipment as shown in **Figure 1** or similar if your light gate equipment is different.

Figure 1

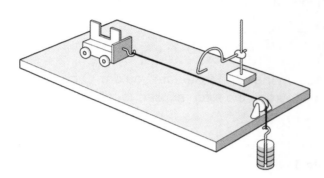

2. Measure the width of each segment of the interrupt car and enter the values into the datalogging software.

3. Place 400 g of mass on the trolley. Add 100 g of mass (1 N of force) to the hook and hold the trolley in place at the 'start line'.

4. Record the acceleration of the trolley as it passes through the light gate using the datalogger. Record this result in **Table 1** using the following headings.

Force (N)	Acceleration (m/s²)			
	First reading	Second reading	Third reading	Mean

Repeat twice more, record your results and calculate the mean acceleration.

5. Take a 100 g mass **off the** trolley and add it to the hook so there is now a force of 2 N. Measure the acceleration for this force.

6. Repeat step **4** up to 5 N of force so all five 100 g masses are on the hook and there are no masses on the trolley.

Method 2 – Using stopwatches

1. Set up the equipment as shown in **Figure 2**.

Figure 2

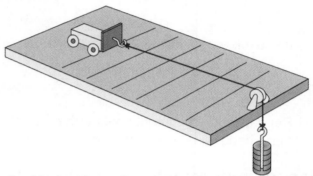

2. Mark at least five straight lines across the surface. They should be an equal distance apart (e.g. 20cm).

3. Place 300 g of mass on the trolley. Add 100 g of mass (1 N of force) to the hook and hold the trolley at the start line.

4. Start the stopwatch as you release the trolley and press 'lap' as the trolley passes each line on the surface. Record the times at each distance for the applied force of 1 N in **Table 1** using the following headings.

Distance travelled (cm)	Time (s)			
	1N	2N	3N	4N

5. Take a 100 g mass **off the** trolley and add it to the hook so the applied force is now 2 N. Measure the time taken for each segment again.

6. Repeat step **4** up to 4 N of force so all four 100 g masses are on the hook and there are no masses on the trolley.

7. Calculate the acceleration of your trolley in each segment, for each force. You could record the calculated acceleration in **Table 1** in a different colour next to the time at each distance.

Activity 2 – Measuring the effect of mass on acceleration with a constant force

1. Set up the equipment as you did in **Activity 1**,

2. Using your results from **Activity 1,** choose an appropriate number of masses to hang on the string to accelerate the trolley along the track.

3. Place 400 g of mass on the trolley. Add your chosen mass to the hook and hold the trolley in place at the 'start line'.

4. Repeat your experiment as you did in **Activity 1** and record your results in **Table 2.**

    If you followed **Method 1** in **Activity 1**, use the following headings for **Table 2.**

Mass of car (g)	Acceleration (m/s²)			
	First reading	Second reading	Third reading	Mean

    If you followed **Method 2** in **Activity 1**, use the following headings for **Table 2.**

Distance travelled (cm)	Mass of car (g)		

5. Repeat your experiment again with fewer masses on the car. Make sure you don't change the mass hanging on the string. This is kept constant to keep the force applied to the trolley constant.

Record your results

Table 1

Table 2

Check your understanding

1. State the independent and dependent variables in Activity 1, and in Activity 2. [3 marks]

   Independent variable ....

   Dependent variable ....

2. **a.** List two variables that were controlled in Activity 1. [2 marks]

   ....

   ....

   b. Explain why it is important that a pulley is used as the string runs over the edge of the bench. [1 mark]

   ....

   ....

Exam-style questions

1. A student investigated the effect of changing the force on the acceleration of a trolley by adding masses to a string attached to a trolley. As the masses fell, the trolley was pulled forward and the student used two light gates to measure the trolley's speed at two points and the time taken to go between the two points. The student used this data to calculate the acceleration.

   The results are shown in **Table 2**.

   Table 2

Force (N)	Acceleration (m/s^2)
1	0.5
2	0.9
3	1.4
4	2.1
5	2.5
6	3.0
7	3.6
8	4.1

   a. Plot the results for force against acceleration on **Graph 1**. [3 marks]

Graph 1

b. Describe the relationship between force and acceleration. [2 marks]

....

....

c. Use the graph to find the acceleration if 3.5 N of force was applied to the trolley. [1 mark]

....

d. **HT** Inertial mass is a measure of how difficult it is to change the velocity of an object.

Use the results of the experiment to calculate the inertial mass of the trolley. Give your answer to 1 significant figure. [3 marks]

....

....

....Inertial mass of the trolley = ... kg

By careful observation and measurement we can measure and calculate the **wavelength** and **frequency** of the waves and then work out their **speed**. We can use a piece of equipment called a ripple tank to explore waves thorugh water. A strobe light can be used to 'freeze' the movement of the waves for making certain measurements. You are going to investigate how to measure the speed of water waves and the speed of waves along a stretched string.

Learning outcomes	Maths skills required	Formulae
• Carry out an investigation to measure the speed of a wave. • Evaluate the method. • Understand how to identify any anomalous results.	• Substitute numerical values into algebraic equations using appropriate units for physical quantities.	• wave speed = frequency × wavelength

Apparatus list

Activity 1
- ripple tank
- strobe light
- low-voltage power pack
- A 5 W or 6 W signal generator
- wooden bridge
- vibration generator

Activity 2
- stretched string or elastic cord
- 100 g and 10 g masses and hangers
- pulley on clamp
- stopwatch
- metre ruler
- eye protection

Safety notes

- Do not handle the power pack, plug or socket with wet hands.
- Be careful with water and electrical pieces of equipment.
- Let your teacher know if you will be affected by stroboscopic or flashing lights.

Common mistakes

- If you are struggling to measure the length of water waves in Activity 1, use a camera or phone to take a photo of the wave patterns next to a ruler.
- If you are struggling to produce a stable wave pattern in the stretched string or elastic, ask your teacher for help.
- In Activity 2 the distance between two points on the string where the vibration amplitude is zero is *half* a wavelength, so the wavelength is *twice* this distance.

Method

Read these instructions carefully before you start work.

There are two activities for you to complete.

Activity 1 – Wave speed through a liquid

1. Set up the ripple tank as shown in **Figure 1**.

2. Count the number of waves that pass a given point every 10 seconds. Record the results in **Table 1**.

3. Measure the length of 10 waves. Record the results in **Table 1**.

4. Change the frequency and take the measurements again.

5. Repeat this until you have at least six sets of results. Find the wave frequency and wavelength for each set of results.

6. Calculate the speed of the waves for each set of results using the equation:
wave speed = frequency × wavelength

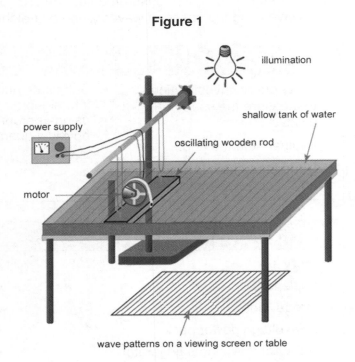

Figure 1

illumination

shallow tank of water

power supply

oscillating wooden rod

motor

wave patterns on a viewing screen or table

Activity 2 – Wave speed through a solid

1. Set up the equipment as shown in **Figure 2**.

2. Put on your eye protection. Turn on the vibration generator. The attached string (or elastic) will start to vibrate up and down.

3. If you cannot see a stable wave pattern along the string, adjust the length of the string which vibrates by slowly moving the wooden bridge away from the vibration generator until it looks like the waves are not moving and you can see clear standing waves. If you cannot see any waves, add another mass to the hanger and then try moving the wooden bridge again.

4. Measure the length of one **wavelength**. Record this in **Table 2**.

5. Record the **frequency** of the signal generator.

6. Calculate the speed of the wave using the equation: wave speed = frequency × wavelength

7. Change the frequency on the signal generator and repeat Steps **3 – 6** twice more.

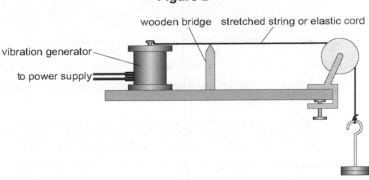

Figure 2

wooden bridge   stretched string or elastic cord

vibration generator

to power supply

Record your results

Table 1 – Calculating wave speed in Activity 1

Result	Number of waves in 10 seconds	Wave frequency (number of waves in 1 second)	Length of 10 waves (m)	Wavelength (length of 1 wave) (m)	Speed (m/s)
1					
2					
3					
4					
5					
6					

Table 2 – Calculating wave speed in Activity 2

Wavelength (m)	Frequency (Hz)	Speed (m/s)

Check your understanding

1. The speed of waves through a liquid should be constant for the same liquid depth.

   a. Which, if any, of your results in Table 1 looks most like it might be an anomalous result?

   [1 mark]

   ....

   b. Calculate the mean wave speed for your wave speeds in Table 1.

   Do not include any anomalous results. [1 mark]

   ....

   Mean wave speed = ... m/s

2. Explain why the method in Activity 1 asks you to measure the length of 10 waves to calculate the wavelength, rather than just measuring the length of one wave. [1 mark]

   ....

3. Identify one possible source of error in the measurement of wavelength in Activity 2. [1 mark]

....

....

Exam-style questions

1. State the units of wave speed, frequency and wavelength. [2 marks]

   a. Wave speed ....

   b. Frequency ....

   c. Wavelength ....

2. State the equation that links wave speed, frequency and wavelength. [1 mark]

....

All objects give off infrared radiation and the warmer it is, the more quickly it emits radiation. However, the colour of an object also affects how much radiation it emits. If we want a radiator to work as well as possible, the colour matters.

You will investigate how the nature of a surface affects the amount of infrared radiation emitted or absorbed.

Learning outcomes	Maths skills required
• Identify independent, dependent and control variables in an investigation. • Carry out a practical investigation to find out how the nature of a surface affects the amount of infrared radiation it emits. • Measure temperature accurately.	• Display experimental data in a suitable table and chart.

Apparatus list

- Leslie cube
- thermometer
- heatproof mat
- ruler
- kettle full of water
- infrared detector

- three test tubes; one painted black, one painted white, one covered in foil
- three thermometers, one in each tube
- test tube rack
- a 250 W bulb and power source
- 10 cm^3 measuring cylinder
- stopwatch

Safety notes

- Be careful with the hot water! Make sure you do not carry containers of hot water across the lab. Take the kettle to your desk and boil the water there if possible.
- **Don't** pick up the Leslie cube when it is full of hot water. It will be very, very hot.
- Do not touch the bulb! It will be very hot even after it has been turned off.

Common mistakes

- In Activity 1, you should keep the sensor the same distance from each face of the cube.
- In Activity 2, you should keep the light bulb the same distance from each test tube.

Method

Read these instructions carefully before you start work.

There are two activities to complete.

Activity 1 – Investigating the amount of infrared radiation radiated from different surfaces

1. Place the Leslie cube onto a heatproof mat.

2. Fill the cube with very hot water and replace the lid of the cube.

3. **DO NOT TOUCH THE LESLIE CUBE.**

4. Place the infrared detector 5 cm from the surface, as shown in **Figure 1**, and measure the amount of infrared radiation radiated from each surface.

5. Measure the temperature using the thermometer. Record your results in **Table 1**.

6. Measure the amount of radiation emitted by the other surface types. Work quickly so the temperature of the water is the same for each measurement.

7. Draw a bar graph of your results in **Graph 1**.

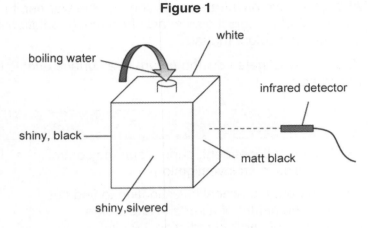

Figure 1

boiling water

white

shiny, black

shiny,silvered

infrared detector

matt black

Activity 2 – Investigating the amount of infrared radiation absorbed by different surfaces

1. Measure 10 cm³ of cold water using the measuring cylinder and pour into the black painted test tube. Repeat for the white painted test tube and the one wrapped in aluminium foil.

2. Place the bulb 5-10 cm away from the test tubes, and switch it on.

3. Take the temperature of the water, record in **Table 2** and start the stop watch.

4. After 5 minutes, take the temperature of the water again and record the temperature in **Table 2**. Repeat every 2 minutes up to 10 minutes.

Record your results

Table 1 – Amount of infrared radiation emitted

Surface type	Temperature of water (°C)	Reading on infrared detector

Table 2 – Effect of coloured surface on absorption of infrared radiation

Surface	Temperature of water (°C)					
	0 mins	2 mins	4 mins	6 mins	8 mins	10 mins
Black paint						
White paint						
Aluminium foil						

Graph 1

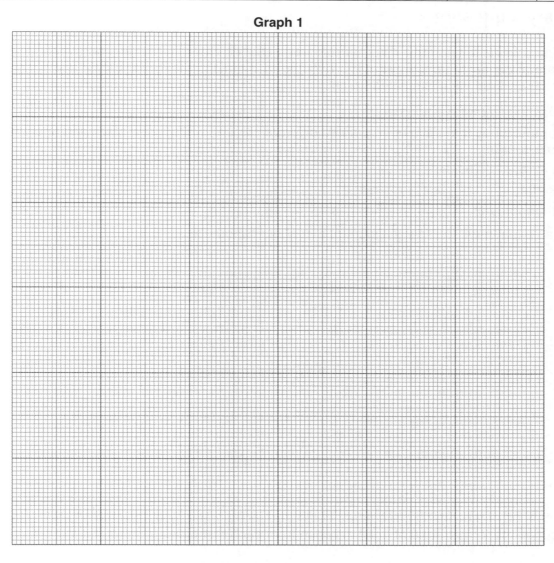

Check your understanding

1. State the independent variable and the dependent variable in **Activity 1**. [2 marks]

Independent variable ....

Dependent variable ....

2. List two control variables in **Activity 1**.   [2 marks]

....

....

3. Describe how you could improve the accuracy of **Activity 2**.   [2 marks]

....

....

Exam-style questions

1. A student wants to find out if different coloured surfaces emit different amounts of radiation when they are at the same temperature. Her hypothesis is that the darker the colour of a surface, the more infrared radiation is emitted from it.

She has the following equipment:
- four test tubes that have been painted different colours: white, grey, black and unpainted (as a control)
- an infrared detector
- a kettle
- a thermometer.

Plan an experiment to allow the student to test her hypothesis.   [6 marks]

....

....

....

....

....

....

....

2. As a method of measuring the amount of infrared radiation emitted, the student suggests measuring the temperature at a set distance from each test tube.

Suggest a reason why this may not be as accurate as using an infrared detector.   [1 mark]

....

....

3. The student's results show that the material the test tube is covered in does affect the amount of radiation emitted. Suggest why.   [1 mark]

.... s

Equations

It is important to be able to recall and apply the following equations using standard units:

Equation number	Word equation	Symbol equation
1	weight = mass × gravitational field strength	$W = mg$
2	work done = force × distance (along the line of action of the force)	$W = Fs$
3	force applied to a spring = spring constant × extension	$F = ke$
4	distance travelled = speed × time	$s = vt$
5	acceleration = $\dfrac{\text{change in velocity}}{\text{time taken}}$	$a = \dfrac{\Delta v}{t}$
6	resultant force = mass × acceleration	$F = ma$
7	momentum = mass × velocity	$p = mv$
8	kinetic energy = 0.5 × mass × (speed)^2	$E_k = \frac{1}{2}mv^2$
9	gravitational potential energy = mass × gravitational field strength × height	$E_p = mgh$
10	power = $\dfrac{\text{energy transferred}}{\text{time}}$	$P = \dfrac{E}{t}$
11	power = $\dfrac{\text{work done}}{\text{time}}$	$P = \dfrac{W}{t}$
12	efficiency = $\dfrac{\text{useful output energy transfer}}{\text{total input energy transfer}}$	
13	efficiency = $\dfrac{\text{useful power output}}{\text{total power input}}$	
14	wave speed = frequency × wavelength	$v = f\lambda$
15	charge flow = current × time	$Q = It$
16	potential difference = current × resistance	$V = IR$
17	power = potential difference × current	$P = VI$
18	power = (current)^2 × resistance	$P = I^2 R$
19	energy transferred = power × time	$E = Pt$
20	energy transferred = charge flow × potential difference	$E = QV$
21	density = $\dfrac{\text{mass}}{\text{volume}}$	$\rho = \dfrac{m}{v}$

The following equations will appear on the equations sheet that you are given in the exam. It is important to be able to select and apply the appropriate equation to answer a question correctly.

Equation number	Word equation	Symbol equation
1	(final velocity)^2 – (initial velocity)^2 = 2 × acceleration × distance	$v^2 - u^2 = 2as$
2	elastic potential energy = 0.5 × spring constant × (extension)^2	$E_e = \frac{1}{2}ke^2$
3	change in thermal energy = mass × specific heat capacity × temperature change	$\Delta E = mc\Delta\theta$
4	period = $\dfrac{1}{\text{frequency}}$	
5	force on a conductor (at right angles to a magnetic field) = magnetic flux density × current × length	$F = BIl$
6	thermal energy for a change of state = mass × specific latent heat	$E = mL$
7	potential difference across primary coil × current in primary coil = potential difference across secondary coil × current in secondary coil	$V_sI_s = V_pI_p$

Answers

4.1.1.2 Microscopy

Check your understanding

1. Use your own results and the following equation:

$$\text{magnification} = \frac{\text{size of image}}{\text{size of real object}}$$

   accurate measurement [1]
   rearranging and using equation [1]
2. With a scale bar, it is always possible to see the size of the cell/object even if the picture is increased or decreased in size (e.g. by a photocopier). [1]
3. It is easier to locate the specimen when you start at a lower magnification; you can then increase the magnification to see more detail. [1]

Exam-style questions

1. **a.** 3.72×10^{13} [1]
   b. 7.8 μm [1]
2. **a.** Any three from: [3]
   - Place the object on a microscope slide.
   - Place a cover slip on top.
   - View the object under the microscope at low magnification (e.g. ×40).
   - Draw the object and label any structures.
   - Increase the magnification (e.g. ×400).
   - Draw your object at higher magnification and label any structures.
   - Draw a suitable scale bar for the images.
   b. Measure the size of the image. [1]
   Divide the size of the real image by the magnification. [1]

4.1.3.2 Osmosis

Check your understanding

1. The water in the solution could evaporate, [1]
   which would make the solution more concentrated. [1]
2. Repeat each reading at least twice more. [1]
   Calculate a mean for each reading. [1]
3. **HT** Your own result: the value where the line of best fit crosses the *x*-axis. [1]

Exam-style questions

1. **a.** $(0.06/1.40) \times 100$ [1]
   percentage change in mass = 4.3% [1]
   b. The potato chips had different masses at the start. [1]

Percentage change shows how much water has been absorbed relative to the mass of the sample. [1]
   c. The student could plot the points on a graph of concentration against % change; [1]
   then draw a straight line of best fit [1]
   and use that point to estimate the isotonic point (or where it crosses the *x*-axis) (and so the concentration of the solution inside the cells). [1]
   OR
   The student could repeat the experiment using smaller intervals between 0.4 mol/dm³ and 0.6 mol/dm³. [1]
   The isotonic point will be close to 0 change in mass. [1]
   Then repeat with even smaller intervals until **no** change in mass is measured. [1]

4.2.2.1 Food tests

Check your understanding

1. Any two from: [2]
   - Ethanol is flammable and could catch fire.
   - Boiling water could cause burns.
   - Reagents could get into eyes/on skin.
   - People could have food allergies.
   - Any other reasonable risk (check with your teacher).
2. Benedict's — lipids
   biuret — proteins
   ethanol — glucose (a simple sugar)
   iodine —————— starch (a complex carbohydrate)

   All four correct [3]
   2 marks for three correct; 1 mark for two or one correct; delete a mark each time more than one line is used from either a reagent or a nutrient.
3. Any one from: [1]
   - pipette
   - 10 cm³ measuring cylinder (volume must be specified – just 'measuring cylinder' is not enough)
   - syringe
   - burette

Exam-style questions

1. **a.** Food sample **A**: protein, lipid [1]
   b. Food sample **B**: starch, glucose [1]
   c. Food Sample **C**: glucose, protein [1]
2. Any food that contains fat, e.g. butter, cheese, cream, oil, nut butters, fatty meat, avocado, etc. [1]

4.2.2.1 Enzymes

Check your understanding

1. **a.** your own result: the pH where the solution turns from black to brown fastest [likely to be between pH 5 and pH 7] [1]

b. your own results: the pH where there is no change from black to brown [you might have two answers here, one for too acidic and one for too alkaline] [1]

2. Any two from: [2]
   - temperature
   - volume of starch solution
   - volume of amylase
   - volume of buffer solution
   - volume of iodine
   - interval between times
   - any other valid control variable.

Exam-style questions

1. **a.** The pH will rise [1]
   as fatty acids are made. [1]
   b. Answer should include: [6]
   - measure an amount of a solution of fats (any named fat, e.g. milk, oil)
   - measure an amount of lipase
   - record the temperature
   - add lipase to the fat solution and start the timer
   - every 30 seconds, measure the pH
   - with a pH probe / universal indicator / any appropriate indicator to test for acids
   - repeat at different temperatures
   - heated by a water bath / hot water in a beaker
   - cooled by an ice bath.
   c. Test a range of pH values from 7 to 14. [1]
   Find the one where lipase breaks down lipids fastest. [1]

4.4.1.2 Photosynthesis

Check your understanding

1. **a.** As light intensity increases, the rate of photosynthesis also increases. [1]
   b. Photosynthesis needs light. The more light there is, the greater the rate of photosynthesis. [1]
2. any control variable [1]
   a valid suggestion for how it could be controlled [1]
   Answer could include:
   - the wavelength of light – use the same light bulb
   - the temperature of the water – use an LED bulb (to release less heat) / use a beaker of water in front of the plant to absorb any thermal energy but allow light to pass through
   - the mass of the plant – use the same piece of plant
   - any valid control variable and suggestion.

Exam-style questions

1. Your answer should include the following points:

   Method [up to 4 marks]
   - Place the plant in the conical flask of water / sodium hydrogencarbonate solution.
   - Place the flask in front of a light source.
   - Attach the gas syringe to the conical flask.
   - Leave for a period of time.
   - Return and record the amount of gas produced.
   - Repeat the experiment using a different light filter to only let a specific wavelength of light through.

   Control variables [up to 2 marks]
   - Ensure flask is same distance from light source.
   - Allow same period of time for gas to be produced.
   - Ensure flask has same sized piece of pondweed.
   - Use the same bulb if filters are changed.
   - Use the same gas syringe.
2. **HT** either temperature or light intensity [1]
3. **HT** **a.** As one factor increases, the other decreases in proportion. [1]
   HT **b.** See graph below:

   - Both axes labelled correctly. [1]
   - Line goes up (section A to B). [1]
   - Line levels off (section B to C). [1]

4.5.2.1 Reaction time

Check your understanding

1. Any one from: [1]
   - As the number of times practised increases, the reaction time decreases.
   - The fewer the practices, the longer the reaction time.
   - Any other sensible/valid hypothesis involving these two variables.
2. **a.** Repeating the same experiment is impossible as the person would have practised. (You could do it with different hands, but that would not be a fair test, as the variables are not the same.) [1]
   b. Use other activities for the experiment OR repeat the test with different people. [1]
3. First record the reaction times of a person without caffeine/coffee/caffeinated fizzy drink. [1]
   Then record the reaction times of the same person after they have consumed caffeine/coffee/caffeinated fizzy drink. [1]

Exam-style questions

1. **a.** The data shows **two** main things:
   - Caffeine reduces the distances the ruler drops. [1]
   - The distance the ruler dropped decreases with more practice / reaction time decreases with more practice. [1]
   Anomaly: Experiment 1, test number 5 (150 mm) [1]
   b. As in the graph below, the second line should be above the first and sloping down to the right. [1]

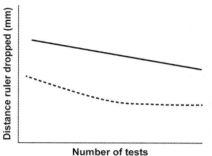

c. There is a limit to how fast humans can react. [1]

4.7.2.1 Field investigations

Check your understanding

1. calculate your answer using:
   (sampled area ÷ total area) × 100 [1]
2. take more samples (e.g. 20 or another named figure) [1]
3. Your answer will depend on your results – you may even have found no relationship. [1]
4. estimate surface cover / percentage cover [1]

Exam-style questions

1. **a.** from the shoreline up the beach / from the sea at low tide to the top of the beach (high tide) [1]
   b. Place the transect and then place a quadrat at equal intervals along the transect. [1]
   Record the number/type of species / surface cover in each quadrat. [1]
   Move the transect along the beach. [1]
   Repeat at least twice more. [1]
2. **a.** 11 (or 12 as daisy on the mid left edge could be included) [1]
   b. mean calculated: 9.8 or 9.9 [1]
   total: 1411 or 1425 daisies (either answer valid) [1]
   If you counted the daisies incorrectly in **2a** but worked out a **correct** mean in **2b** and then multiplied your **correct** answer to **2b** by 144, award yourself both marks for **2b**.
   If you counted the daisies incorrectly in **2a** and worked out a **wrong** mean in **2b**, but then you multiplied your **wrong** answer to **2b** by 144, score 1 mark for **2b**.

5.4.2.3 Making salts

Check your understanding

1. **a.** There is a greater amount of a reactant than necessary to react completely with the limiting reactant. [1]
   b. Harmful sulfur dioxide gas could be produced when the mixture is heated. [1]
2. **a.** A_r Cu = 63.5
   A_r O = 16
   M_r CuO = 63.5 + 16 = 79.5 [1]
   for calculating M_r of CuO

   A_r H = 1
   A_r S = 32
   A_r O = 16
   M_r H$_2$SO$_4$ = (1 x 2) + 32 + (16 x 4) = 98
   [1] for calculating M_r of H$_2$SO$_4$

M_r reactants = 79.5 + 98 = 177.5 [1]
for adding M_r of CuO and M_r of H$_2$SO$_4$.
(Allow mark for error from previous M_r calculations carried forward.)

b. No atoms are lost or made during a chemical reaction (so the mass of the products equals the mass of the reactants). [1]
c. The mass of the products equals the mass of the reactants. [1]
The total mass of the products should be 177.5, or carry forward student's answer from (**a**). [1]

Exam-style questions

1. **a.** The crystals are copper chloride. [1]
   b. Answer should include;
   - add excess copper oxide to hydrochloric acid
   - stir with a glass rod
   - heat gently in a beaker until the copper oxide is dissolved
   - filter to remove unreacted copper oxide
   - place the solution in an evaporating basin over a beaker acting as a water bath
   - evaporate at least half the liquid
   - check for the crystallisation point using a cool glass rod
   - pour the solution into a petri dish
   - allow it to crystallise. [1 each]
2. **a.** CaCO$_3$ + HCl + $\longrightarrow$ CaCl$_2$ + H$_2$O + CO$_2$ [1]
   b. calcium chloride [1]
3. **a.** sodium and nitric acid [1]
   b. sodium is very reactive/it could explode [1]

5.4.3.4 Electrolysis

1. copper at the cathode/negative electrode
   chlorine gas at the anode/positive electrode [1, both required]
2. **a.** CuCl$_2$ [1]
   b. 2 [1]
   c. reduced [1]
3. **a.** Use a burning splint held at the open end of a test tube of the gas. Hydrogen burns rapidly with a pop sound. [1]
   Use damp litmus paper. When damp litmus paper is put into chlorine gas, the litmus paper is bleached and turns white. [1]
   b. Any two from:
   - wear eye protection
   - use a low potential difference (to make sure excessive chlorine is not produced)
   - do not run electrolysis for more than five minutes (to make sure excessive chlorine is not produced)
   - any other sensible precaution. [2]

Exam-style questions

1. **a.** chlorine (anode) [1]
   hydrogen (cathode) [1]
   b. Iodine would be formed at the anode (as I⁻ is a negative ion). [1]
   Hydrogen would be formed at the cathode (as H⁺ is a positive ion). [1]
   Hydrogen is made rather than potassium

because potassium is more reactive than hydrogen, so K^+ ion would stay in solution. [1]
Iodine is made rather than oxygen as iodine is a halogen (group 7) so stays in solution. [1]

   c. potassium hydroxide [1]

2. a. At the negative electrode/anode, a Cu^{2+} ion is reduced and becomes a Cu atom. [1]
$Cu^{2+} + 2e^- \rightarrow Cu$ [1]

   b. When solid, bromine and lead ions are held in a lattice and cannot move. [1]
When molten, the ions can move. [1]

   c. Aqueous lead bromide is dissolved in water. [1]
There are hydrogen ions in water. [1]
Hydrogen is lower than lead in the reactivity series. [1]

5.5.1.1 Temperature changes

Check your understanding

1. a. It is important as otherwise the temperature rise will be lower than the true value OR the experiment will not be accurate. [1]

   b. By adding a lid to the polystyrene cup OR stirring OR any other sensible suggestion. [1]

2. a. $0.1cm^3$ [1]

   b. Repeats can be used to discount anomalous results OR repeats can be used to calculate a mean. (It is not enough to say that repeats might make it more accurate.) [1]

Exam-style questions

1. a. $(101 + 102)/2$ [1]
$= 102$ (s) [1]
If you didn't spot the anomaly and calculated 105 [1]

   b. Any one from: the results are all similar / close together / spread over a small range. [1]

   c. Either:
   • the more concentrated the acid, the shorter the time taken for the magnesium to disappear OR
   • the more dilute the acid, the longer the time taken for the magnesium to disappear. [2]

   d. The particles are closer together in an acid with a higher concentration OR there are more particles in a smaller volume. [1]
More particles are more likely to collide (so reactions are more likely to occur). [1]

2. **HT** The energy required to break the bonds of the reactants is less than the energy released when the bonds of the products are formed. [1]
The net energy change is a transfer of energy to the environment. [1]

5.6.1.2 Rates of reaction

Check your understanding

1. a. As concentration increases, time taken to react decreases. [1]
As there are more particles per unit volume because of the increase in concentration, the frequency of collision increases. [1]

   b. One from:
   • temperature
   • addition of catalyst
   • surface area
   • pressure. [1]

2. You could repeat the experiment at different concentrations. [1]

Exam-style questions

1. a. As temperature increases, rate of reaction increases. [1]
Increasing the temperature increases the frequency of collisions and makes the collisions more energetic. [1]
There are more successful collisions / more collisions have the activation energy. [1]

   b. Answer should include;
   • measure out the same amount of sodium thiosulfate using a **measuring cylinder**
   • heat/cool to different temperature by adding to **ice bath** and **water bath**
   • measure the temperature using a **thermometer**
   • add an amount of hydrochloric acid to the sodium thiosulfate
   • start the **stopclock** as the acid is added
   • time until the **black cross** cannot be seen
   • repeat this experiment for different temperatures
   • plot the results on a graph
   • draw a tangent to the graph and calculate the gradient
   • rate of reaction = 1/time taken for cross to disappear. [up to 6]

2. a. $2 H_2O_2 \rightarrow 2 H_2O + O_2$
1 mark for O_2 [1]
1 mark for correct balancing [1]

   b. Catalysts increase rate of reaction by providing a different pathway for the reaction [1]
that has a lower activation energy. [1]

   c. 1 mark for a valid way to measure rate (e.g. gas production or mass lost) [1]
1 mark for repeating with catalyst added [1]

5.8.1.3 Chromatography

Check your understanding

1. a. near the solvent front [1]

   b. This substance is very soluble in this solvent OR it spends more time in the mobile phase (dissolved in the solvent) than it does in the stationary phase (bound to the chromatography paper). [1]

2. a. near the pencil line [1]

   b. This substance is not at all soluble OR it has spent all of the time in the stationary phase (bound to the chromatography paper) and no time in the mobile phase. [1]

3. The solvent could evaporate. [1]

4. The student could either use pens that are soluble in water [1]
or they could change the solvent for one that is appropriate for permanent markers, i.e. one that permanent markers will dissolve in, such as alcohol. [1]

Exam-style questions

1. **a.** phenylalanine [1]
   b. phenylalanine [1]
2. **a.** A substance cannot travel further than the solvent front OR the distance travelled by the spot/the distance travelled by the substance cannot be greater than 1. [1]
   b. Allow answers between the following ranges:
   spot 1: R_f value of 0.10 – 0.25 [1]
   spot 2: R_f value of 0.50 – 0.65 [1]
   spot 3: R_f value of 0.70 – 0.85 [1]
   c. E102 [1]
   E133 [1]
   d. Test a large range of samples of known E number food additives. [1]
   Compare this to the unknown sample on the chromatogram. [1]
   Calculate the R_f values and match them. [1]
   (If no other mark awarded, give the following 1 mark:)
   Look up known R_f value for E numbers [1]

5.10.1.2 Water purification

Check your understanding

1. Any valid risk and control method can be accepted examples of expected answers are:

Risk	Control method
Using a Bunsen burner and hot equipment	Use tongs for hot equipment
	Ensure any loose clothing and long hair tied back
The distillate/distilled water getting sucked back in to the hot conical flask causing it to crack	Heat the conical flask constantly
	Make sure the delivery tube is above the level of the distillate/distilled water
Getting salt water in eyes	Wear eye protection

2. More samples should be taken [1]
   from different locations. [1]
   Repeats should be conducted (and a mean should be calculated). [1]
3. For an entire community, distilling sea water would require a vast amount of energy [1]
   and would take a long time. [1]

Exam-style questions

1. **a.** 11.06 – 10.97 = 0.09 (g) [1]
   b. (At least) two more repeats could be done [1]
   and a mean could be calculated. [1]
2. **a.** Answer should include;
   - use a conical flask with a bung and delivery tube

- pour the sea water into the conical flask and heat with a Bunsen burner
- until it evaporates
- condense the evaporated water
- using a cold test tube in icy water
- using a (Liebig) condenser
- collect the distillate/distilled water in the test tube/beaker [up to 6]
   b. Potable water contains dissolved substances (but pure water only contains water molecules). [1]

6.1.1.3 Specific heat capacity

Check your understanding

1. to ensure that the entire mass of the water is heated evenly [1]
2. **a.** 69 °C (69.3 °C also allowed) [2]
   b. Any one from: [1]
   - misreading of thermometer
   - fault with equipment
3. The thermal energy from the heating element may be transferred to the surrounding air rather than to the block of metal. [1]
4. balance (accept scale) [1]
   measuring cylinder [1]
   thermometer [1]

Exam-style questions

1. **a.** Your graph should look something like this:

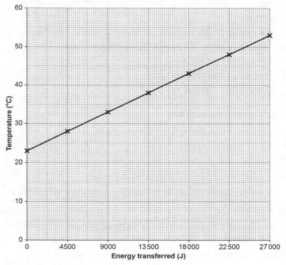

1 mark for plotting 1 or 2 correct points [1]
1 mark for all 3 points correctly plotted [1]
1 mark for line of best fit [1]

   b. gradient of line = 30/27 000 = 0.0011 [1]
   (allow a mark for any other set of values that give a gradient of 0.0011)
   specific heat capacity = 1/(1 × 0.0011)
   = 909.1 [1]
2. Copper would heat up the fastest. [1]
   As it has the lowest specific heat capacity it needs less energy to raise 1 kg by 1 °C. [1]

6.2.1.3 Resistance

Check your understanding

1. As the length of the wire increases, the resistance also increases. [1]
2. The relationship is directly proportional [1] as it is a straight line that passes through the origin. [1]
3. If the resistance is higher, then the current will be lower. [1]
   A lower current passing through a wire will mean it doesn't get as hot / less energy will be transferred as thermal energy. [1]

Exam-style questions

1. **a.** Potential difference (V) = V (volts)
   b. Current (I) = A (amperes, amps)
   c. Resistance (R) = Ω (ohms) [2]
      2 marks for all three correct;
      1 mark for two or one correct
2. $V = IR$
   resistance = potential difference ÷ current
   1 mark for either version [1]
3. As resistance increases, current decreases. [1]

6.2.1.4 I–V characteristics

Check your understanding

1. As potential difference increases, current increases (in both directions). [1]
2. So that the temperature of the resistor is kept constant. [1]
   Resistance increases as temperature increases. [1]
3. Use any pair of values from results table. [1]
   Calculate using $R = V/I$ [1]
4. Repeat each reading. [1]
   Calculate a mean. [1]

Exam-style questions

1. correct substitution into formula: 6.6/0.3 [1]
   resistance = 22 Ω [1]
2. Answer should include: [6]
   - Ammeter used to measure current.
   - Voltmeter used to measure potential difference.
   - Resistance of variable resistor altered to change current in circuit or change potential difference (across diode).
   - Plot a graph of current against potential difference.
   - Resistance (of diode) calculated from pairs of points on the graph, for current in both directions using R = V/I
   - Resistance calculated for a large enough range of currents, flowing in both directions.

6.3.1.1 Density

Check your understanding

1. **a.** Any one of the following points:
   - Air could be caught in crevices of a particular shape, displacing more volume.
   - A mean calculated from repeats makes the results more accurate / closer to the true value.

- Repeats reduce the effect of random error.
- Anomalous values can be identified and discarded.
   b. The density will be lower. [1]
   The air bubble will cause an increase in volume of the modelling clay blob but not an increase in mass (or negligible increase in mass). [1]
2. **a.** Any one of the following:
   - a ruler
   - Vernier callipers
   - (a large Eureka can) [1]
   b. Either of the following points:
   - Measure one side of the cube and calculate (side)³ or (just in case it isn't a regular cube), multiply the base by the width by the height.
   - (The cube could be submerged in the Eureka can and the volume of water displaced could be measured.) [1]

Exam-style questions

1. 1 mark for each of the following points:
   - Measure the volume of the petrol (e.g. by using a measuring cylinder). [1]
   - Measure the mass of the petrol (e.g. using a digital balance). [1]
   - Account for the mass of the container; measure mass of container and take away from mass of liquid + container. [1]
   - Use the formula: density = mass ÷ volume [1]
2. Density of aluminium = 2700 kg/m³ [2]
   5 400 000 g = 5400 kg
   5400 ÷ 2.0 = 2700
   2 marks for correct answer; 1 mark for incorrect answer caused by incorrect unit conversion
3. Mass of steel block = 15 400 kg [2]
   Using formula: density = mass ÷ volume, or mass = density × volume
   2.0 × 7700 = 15 400
   2 marks for correct answer; 1 mark for incorrect answer caused by incorrect formula rearrangement
4. Aluminium has a lower density than steel. [1]
   Aluminium is better as it has less mass per cubic metre and so less energy (fuel, force, lift) is needed to raise each cubic metre off the ground. [1]
   (Answer must refer to density – just saying 'because its lighter' or similar is not enough.)

6.5.3 Force and extension

Check your understanding

1. As force increases, extension also increases. [1]
   The relationship is (directly) proportional. [1]
2. Your own result – the point on the graph at which relationship is no longer directly proportional. [1]
3. Repeat the experiment and calculate the mean. [1]
4. Your answer will depend on your results – correctly calculated for a pair of values when the relationship is directly proportional or calculated using gradient of the graph (units are not required). [1]

Exam-style questions

1. **a.** 8 cm [1]
   b. force = spring constant × extension [1]

c. The gradient of the graph is 1/spring constant.
The spring constant can be found from
1/gradient
gradient = 4
1 ÷ 4 = 0.25
Or, by using $F = ke$, doe pairs of values, for the part of the graph that is linear
2.0 N ÷ 8.0 cm = 0.25 [1]
spring constant = 0.25 N/cm [1]
= 25 N/m [1]
d. e = 4.0 cm (read from graph) [1]
4.0 cm = 0.04 m [1]
k = 25 N/m (from part **c**)
$E_e = 0.5 \times 25 \times 0.04^2$
$E_e = 0.02$ [1]
unit = J (joules) [1]

6.5.6.2.2 Acceleration

Check your understanding

1. Independent variable in Activity 1 = force
Dependent variable in Activity 1 = acceleration
Independent variable in Activity 2 = mass
Dependent variable in Activity 2 = acceleration [3]
3 marks for all four correct; 2 marks for three correct; 1 mark for two or one correct
2. a. Any two from: [2]
• mass of the system being accelerated
• same surface friction
• same distance between velocity measurements (length of U shaped card)
• any other control variable
b. EITHER to reduce (effects of) friction OR to keep the trolley moving a straight line. [1]

Exam-style questions

1. a. See graph below:

both axes labelled correctly [1]
points plotted accurately [1]
line of best fit drawn [1]
Subtract 1 mark if axes are reversed – dependent variable should be plotted on *y*-axis
b. As force increases, acceleration increases. [1]
The relationship is (directly) proportional. [1]
c. 1.75 m/s² (units are required) [1]
(anything between 1.5 m/s² and 2.0 m/s² is acceptable)
d. **HT** EITHER use $F = ma$
Take any pair of values from the table for force ÷ acceleration, showing working [1]
answers between 0.45 and 0.53 [1]
rounded to 0.5 kg (1 significant figure) [1]
OR

calculate the gradient of the graph [1]
gradient between 0.45 and 0.53 [1]
rounded to 0.5 kg (1 significant figure) [1]

6.6.1.2 Waves

Check your understanding

1. a. Your answer will depend on your findings – you may notice that one result is far from the rest of the spread of results. [1]
b. Your own results. Calculate the sum of all the wave speeds, excluding any anomalies, and then divide by the total number of results. [1]
2. EITHER it is easier to measure a larger distance accurately (i.e. the length of 10 waves) than it is to measure one shorter wavelength
OR it is harder to define and pick out a single wave than a set of several waves. [1]
3. Any one from: [1]
• The movement of the string is very fast / a blur, so difficult to decide exactly where to measure from.
• Metre ruler not held parallel to the string, or ruler itself is not straight.
• Parallax error when measuring the length of string with the metre ruler.
• Not converting the distance measured to a wavelength (distance could be 1, 2, 3, 4 or 5 half wavelengths)

Exam-style questions

1. a. Wave speed (v) = m/s (or metres per second or m s⁻¹)
b. Frequency (f) = Hz (or hertz)
c. Wavelength (λ) = m (or metres)
2 marks for all three correct answers
1 mark for one or two correct answers
correct use of letter case required for all three (i.e. 'M' is incorrect for metres) [2]
2. wave speed = frequency × wavelength
$v = f \times \lambda$ [1]
Word or symbol equation acceptable;
also accept $f = v/\lambda$ or $\lambda = v/f$

6.6.2.2 Radiation and absorption

Check your understanding

1. Independent variable = colour of surface [1]
Dependent variable = amount of infrared radiation emitted by surface [1]
2. Any two from: [2]
• distance of infrared detector from surface
• starting temperature of water
• amount of water
• same area of surface
• any other control variable.
3. Any two from:
• use a temperature sensor and data logger instead of a glass thermometer to measure temperature [1]
• take repeat readings (using the same starting water temperature) [1]
• calculate a mean (excluding any anomalies) [1]

Exam-style questions

1. Answer should include: [6]
   - Boil a kettle.
   - Measure the same amount of boiling water using a measuring cylinder.
   - Put water in the different test tubes; use test tubes that are the same size and shape.
   - Place the infrared detector by the test tubes, the same distance away.
   - Measure the amount of infrared radiation emitted.
   - Repeat for the different test tubes, working quickly so the temperature of the water is the same for each measurement.
   - Take repeat readings using the same initial water temperature, and find the mean (discarding any anomalous results).

2. Thermometers have a poor resolution compared to infrared sensors, so cannot detect small differences in temperature from the different faces. [1]

3. Different materials emit different amounts of radiation. [1]